Music Projects

The Maplin series

This book is part of an exciting series developed by Butterworth-Heinemann and Maplin Electronics Plc. Books in the series are practical guides which offer electronic constructors and students clear introductions to key topics. Each book is written and compiled by a leading electronics author.

Other books published in the Maplin series include:

Audio IC projects	Maplin	0 7506 2121 4
Computer interfacing	Graham Dixey	0 7506 2123 0
Logic design	Mike Wharton	0 7506 2122 2
Starting Electronics	Keith Brindley	0 7506 2053 6

Music Projects

Robert Penfold

NEW·TECH

Newnes
An imprint of Butterworth-Heinemann Ltd
Linacre House, Jordan Hill, Oxford OX2 8DP

A member of the Reed Elsevier group

OXFORD LONDON BOSTON
MUNICH NEW DELHI SINGAPORE SYDNEY
TOKYO TORONTO WELLINGTON

British Library Cataloguing in Publication Data
A catalogue record for this book is available from the
British Library
ISBN 0 7506 2119 2

Library of Congress Cataloguing in Publication Data
A catalogue record for this book is available from the
Library of Congress

Edited by Co-publications, Loughborough
Typeset and produced by Sylvester North, Sunderland

— all part of The Sylvester Press

Printed in England by Clays Ltd, St Ives plc

Contents

Preface

This book is a collection of projects, specifically for electro-music applications, designed by the well-known electronics author Robert Penfold.

This is just one of the Maplin series of books published by Newnes books covering all aspects of computing and electronics. Others in the series are available from all good bookshops.

Maplin Electronics Plc also publishes a monthly electronics magazine called *Electronics*; it is the ideal choice for anyone who wants to keep up with the world of electronics, computing, science and technology. Practical electronic projects are included with all parts readily available.

Maplin Electronics Plc supplies a wide range of electronics components, project kits, tools, cables and connectors suitable for electro-music applications and other products to private individuals and trade customers. Telephone: (0702) 552911 or write to Maplin Electronics, PO Box 3, Rayleigh, Essex SS6 8LR, for further details of product catalogue and locations of regional stores.

·

1 Guitar effects

Featuring:

Music projects

Fuzz unit

There are numerous types of fuzz unit, giving a range of subtle variations on the basic effect. The most common form of fuzz effect is a straightforward clipping of the signal to introduce severe harmonic distortion. A point which should be borne in mind is that a fuzz unit working on that principle generates strong intermodulation distortion which restricts its use to just one note at a time. This fuzz unit here, on the other hand, is slightly different to the conventional type, but it *is* basically a clipping circuit nevertheless. It is different in that it incorporates an envelope shaper that retains the original attack and decay characteristics of the guitar, or at least something approximating to these. Normally a simple clipping amplifier is used and this gives a virtually constant output level until the signal decays below the point where clipping and the fuzz effect are produced. Of course, in practice the next note is normally produced before the output has a chance to decay that far, and the output lacks any real shape at all.

In this circuit IC1 is an operational amplifier connected in non-inverting mode. The inclusion of D1 and D2 in the negative feedback loop introduces the clipping. At output levels of less than about 1 volt peak-to-peak D1 and D2 are not brought into conduction as their forward threshold voltage of about 0.6 volts is not reached. The voltage gain of the amplifier is consequently controlled by RV1 and R3. At output levels of much more than about 1 volt peak-to-peak D1 and D2 are brought into conduction (D1 on negative peaks and D2 on positive peaks). They shunt RV1 to give more feedback, lower voltage

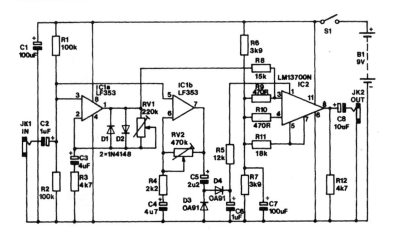

Figure 1.1   Fuzz unit circuit diagram

gain, and the required distortion. RV1 is adjusted so that the basic voltage gain of the amplifier is sufficient for the guitar to produce strong clipping.

The output of IC1(a) is fed to a voltage controlled amplifier (VCA) built around operational transconductance amplifier IC2. This is a conventional circuit which uses one section of the LM13700N (the other section is left unused). The voltage gain of the circuit is controlled by the bias current fed to the amplifier bias input at pin 1, but the addition of series resistor R5 gives a current flow that is roughly proportional to the applied voltage, and voltage rather than current control is consequently obtained. The control voltage is obtained by amplifying, rectifying, and smoothing the input signal. IC1(b) provides the amplification while D3, D4 and C6 provide rectification and smoothing. The control voltage is

3

roughly proportional to the amplitude of the input signal, as is the gain of IC2. The almost constant output level from IC1(a) is therefore shaped to give an output which retains a reasonable approximation to the original envelope shape of the input signal. A useful by-product of the envelope shaping is that a noise gate action is obtained. This avoids the problems of high noise, hum pick-up and feedback that accompany the use of many fuzz units.

RV1 is adjusted to give the required degree of fuzz. A low setting gives little or no distortion while — with most guitars — a high setting produces a strong fuzz effect even when the signal has decayed to a low level. RV2 is adjusted simply by trial and error to find the setting which gives the best envelope shape. With too low a setting the signal decays too early, but with too high a setting the envelope shaping is virtually nonexistent.

Fuzz unit parts list

Resistors — All 0.6 W 1% metal film (unless stated)

R1,2	100 k	2	(M100K)
R3,12	4k7	2	(M4K7)
R4	2k2	1	(M2K2)
R5	12 k	1	(M12K)
R6,7	3k9	2	(M3K9)
R8	15 k	1	(M15K)
R9,10	470 Ω	2	(M470R)
R11	18 k	1	(M18K)
RV1	220 k hor sub-min preset	1	(UH07H)
RV2	470 k hor sub-min preset	1	(UH08J)

Capacitors

C1,7	100 µF 10 V axial electrolytic	2	(FB48C)
C2,6	1 µF 100 V axial electrolytic	2	(FB12N)
C3,4	4µ7F 100 V axial electrolytic	2	(FB18U)
C5	2µ2F 100 V axial electrolyic	1	(FB15R)
C8	10 µF 25 V axial electrolytic	1	(FB22Y)

Semiconductors

IC1	LF353N	1	(WQ31J)
IC2	LM13700N	1	(YH64U)
D1,.2	1N4148	2	(QL80B)
D3,4	OA91	2	(QH72P)

Music projects

Miscellaneous

S1	SPST ultra-min toggle	1	(FH97F)
JK1,2	$^1/_4$ in jack skt brk	2	(HF90X)
B1	9 V PP3-sized battery	1	(FK58N)
	8-Pin DIL socket	1	(BL17T)
	16-Pin DIL socket	1	(BL19V)
	battery clip	1	(HF28F)

Envelope tremolo

Tremolo is perhaps the oldest of electronic musical effects, and works on the principle of simply amplitude modulating the input signal, or varying its volume in other words. With the conventional method the modulation is provided by a low frequency sinewave or triangular waveform generator, with an adjustable frequency range of around 0.5 to 10 Hz. This gives quite a useful effect, but one which sounds rather mechanical due to the lack of any change in modulation rate during the course of each note.

A much more interesting effect can be obtained by having the modulation frequency related to the envelope of the input signal, and it is this approach that has been adopted for this tremolo unit. The modulation frequency starts at a relatively high figure when the guitar note is first struck and the volume is high, and it gradually reduces as the volume of the note decays.

Operational amplifier IC1 acts as an input buffer and amplifier stage. If the unit is to be used with a high output guitar, or some other electronic instrument such as a monophonic synthesiser, resistor R2 should have a value of 47 k so that IC1 acts as a unity voltage gain buffer stage. For use with a low output guitar, R2 should be increased to 470 k so that IC1 provides a voltage gain of about 11 times, and boosts the output of the guitar to an adequate level to drive this unit properly.

The voltage controlled amplifier is based on integrated circuits IC2 and IC3. IC2 is a transconductance operational amplifier which is connected in a standard voltage

Music projects

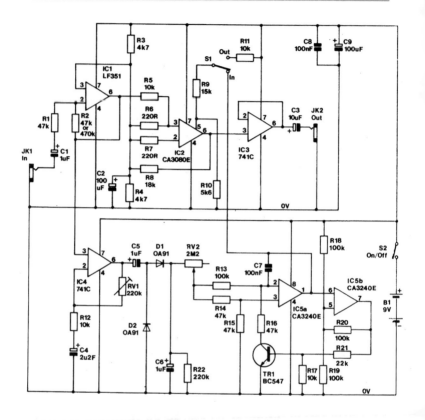

Figure 1.2 Envelope tremolo circuit diagram

controlled amplifier configuration with IC3 acting as the output buffer stage. The bias voltage fed to resistor R9 controls the gain of the voltage controlled amplifer, and with S1 in the out position a fixed bias is provided so that the input signal passes through the unit without any modulation being introduced. When set to the *in* position, switch S1 connects the control input of the voltage controlled amplifer to the output of a voltage controlled oscillator.

The voltage controlled oscillator is built. around integrated circuit IC5, which is used in a circuit that is very similar to an ordinary triangular/squarewave oscillator of the type which is based on a Miller integrator and a Schmitt trigger. In fact, IC5(a) operates as an integrator, and IC5(b) acts as a trigger circuit, but the standard configuration has been modified somewhat to provide voltage controlled operation. The control voltage is applied to the junction of resistors R13 and R14. It is, of course, the triangular output from IC5(a) that is used as the modulation signal, and not the squarewave from IC5(b).

The control signal is generated by first amplifying the output from integrated circuit IC1 using IC4 as a non-inverting amplifier. The output from IC4 is then rectified and smoothed by diodes D1, D2 and D6 to give a d.c. control voltage which is roughly proportional to the amplitude of the input signal. Potentiometer RV2 enables the maximum modulation frequency to be controlled.

The finished unit only requires one adjustment before it is ready for use, and this is to set preset RV1 for a suitable voltage gain through IC4. This is really a matter of trial and error to find a setting that gives good results. If set too low in value, the voltage gain of the circuit will be inadequate, and only very low modulation frequencies will be obtained even with RV2 set at minimum resistance (which corresponds with maximum modulation frequency range). Setting preset RV1 at maximum resistance will result in the modulation frequency remaining virtually constant at a high figure until the note has almost fully decayed. When RV1 is set correctly, a

Music projects

fairly high maximum modulation frequency should be attainable, but the frequency should start to reduce downwards soon after the beginning of each note.

Envelope tremolo parts list

Resistors — All 0.6 W 1% metal film (unless stated)

R1,14–16	47 k	4	(M47K)
R2	See text		
R3,4	4k7	2	(M4K7)
R5,11, 12,17	10 k	4	(M10K)
R6,7	220 Ω	2	(M220R)
R8	18 k	1	(M18K)
R9	15 k	1	(M15K)
R10	5k6	1	(M5K6)
R13,18, 19,20	100 k	4	(M100K)
R21	22 k	1	(M22K)
R22	220 k	1	(M220K)
RV1	220 k sub-min hor preset	1	(UH07H)
RV2	2M2 lin pot	1	(FW09K)

Capacitors

C1,5,6	1 µF 100 V PC electrolytic	3	(FF01B)
C2,9	100 µF 10 V PC electrolytic	2	(FF10L)
C3	10 µF 50 V PC electrolytic	1	(FF04E)
C4	2µ2F 100 V PC electrolytic	1	(FF02C)
C7	100 nF poly layer	1	(WW41U)
C8	100 nF ceramic	1	(BX03D)

Music projects

Semiconductors

IC1	LF351N	1	(WQ30H)
IC2	CA3080E	1	(YH58N)
IC3,4	LM741CN	2	(QL22Y)
IC5	CA3240E	1	(WQ21X)
TR1	BC547	1	(QQ14Q)
D1,2	0A91	2	(QH72P)

Miscellaneous

JK1,2	$^1/_4$ in jack skt brk	2	(HF90X)
S1	press toe switch SPDT	1	(FH92A)
S2	SPST ultra-min toggle	1	(FH97F)
B1	9 V PP3-sized battery	1	(FK58N)
	battery clip	1	(HF28F)
	8-pin DIL socket	5	(BL17T)

Tremolo unit

A tremolo unit is one of the most simple types of musical effects unit, although one would probably not guess this from the prices of ready-made units. The design featured here can be built at low cost but it nevertheless has a high level of performance.

The function of a tremolo unit is to amplitude modulate an input signal, with the modulation frequency being variable from typically about 0.5 to 5 Hz or so. The modulation waveform must be one that has a low harmonic content, such as a sine or triangular type, so that a smooth and pleasant effect is produced. In this circuit the modulation signal is generated by integrated circuit IC2 which is a dual operational amplifier used as a conventional triangular waveform generator; this is a form of relaxation oscillator which uses IC2(a) as a Miller integrator, and IC2(b) as a Schmitt trigger. This gives a squarewave output from IC2(b), and the required triangular waveform from IC2(a). Timing capacitor C3 charges and discharges via resistor R6 and potentiometer RV2, and the operating frequency of the oscillator can therefore be controlled using RV2. This gives an approximate frequency range of 0.5 Hz to 10 Hz.

The modulator uses MOSFET integrated circuit IC1 as a simple voltage controlled resistor. Integrated circuit IC1 is a CMOS 4007UBE device, which contains two complementary pairs and an inverter. In this design, only one (N channel) transistor of one complementary pair is used and the other parts of the device are totally ignored. The drain to source resistance of integrated circuit IC1 forms

Music projects

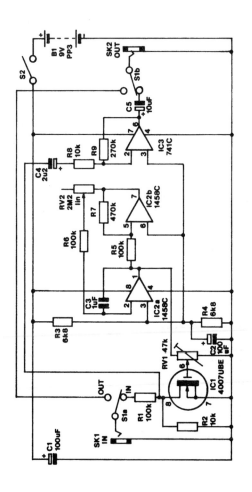

Figure 1.3 Tremolo unit circuit diagram

14

an attenuator in conjunction with resistor R1. However, resistor R2 is connected in parallel with IC1, and this ensures that there is always a loss of 20 dB or more through the attenuator. This is to keep the signal voltage across IC1 at a low level so that good low-distortion performance is obtained. The losses through the attenuator reach a maximum of about 50 dB with IC1 biased into saturation. The gate of integrated circuit IC1 is driven from the output of the modulation oscillator via potentiometer RV1. The latter is adjusted to give an input voltage range to IC1 that gives good symmetrical modulation, and this is really just a matter of adjusting RV1 to obtain what is judged to be the best effect, unless suitable test gear (an audio frequency signal generator and an oscilloscope) is to hand.

Integrated circuit IC3 is used as an amplifier and buffer stage which compensates for the losses through the attenuator and provides the unit with a low output impedance. Switch S1 is a bypass switch, and in practice this is a heavy duty (latching) push button switch mounted on the top panel of the case so that it can be foot operated. The case must be a strong type, such as a diecast aluminium box. The current consumption of the circuit is very low at only about 4 milliamps.

Tremolo unit parts list

Resistors — All 0.6 W 1% metal film (unless stated)

R1,5,6	100 k	3	(M100K)
P,2,8	10 k	2	(M10K)
R3,4	6k8	2	(M6K8)
R7	470 k	1	(M470K)
R9	270 k	1	(M270K)
RV1	47 k sub-min hor preset	1	(UH05F)
RV2	2M2 lin pot	1	(FW09K)

Capacitors

C1,2	100 µF 10 V axial electrolytic	2	(FB48C)
C3	1 µF poly layer	1	(WW53H)
C4	2µ2 100 V axial electrolytic	1	(FB15R)
C5	10 µF 25 V axial electrolytic	1	(FB22Y)

Semiconductors

IC1	HCF4007UBEY	1	(QX04E)
IC2	MC1458CN	1	(QH46A)
IC3	LM741CN	1	(QL22Y)

Miscellaneous

S1	press toe DPDT	1	(FH93B)
S2	SPST ultra min toggle	1	(FH97F)
B1	9 V PP3-sized battery	1	(FK58N)
SK1,2	$^1/_4$ in jack skt brk	2	(HF90X)
	14-pin DIL IC socket	1	(BL18U)

Alternative tremolo unit

This tremolo is also one of the oldest and most simple of electronic music effects. The effect is produced by varying the amplitude of the processed signal at a low frequency (usually around 1 Hz to 2 Hz). It can be generated manually using a swell pedal by operating the pedal at the appropriate rate, but an automatically generated effect is far more convenient in use. This circuit provides a conventional tremolo effect at a rate which is variable from under 1 Hz to around 10 Hz. It is easily modified to provide a more dynamic effect — this is described later.

It is important that the low frequency modulation oscillator provides an appropriate waveform, which means one that provides smooth variations in volume. This basically means either a sinewave or a triangular-wave modulation signal, and in this case it is a triangular waveform that is used. This is simply because a triangular waveform is easier to generate than a reasonably pure sinewave type. The modulation signal is provided by integrated circuit IC1 which is used in a modified version of the standard squarewave/triangular waveform generator. It is actually a form of voltage controlled oscillator (VCO), having the control voltage provided by potentiometer VR1. The latter therefore acts as the modulation rate control.

The amplitude modulation is added to the input signal via a simple voltage controlled amplifier based on integrated circuit IC2. This is a CMOS inverter and a complementary pair of MOSFET transistors, but in this case only one N-channel MOSFET transistor of this device is actually used. Thus connections are only made to three pins of IC2, and the other eleven are just ignored. The transistors in the CMOS integrated circuit are enhancement mode devices, which means that they are normally switched off and are biased into conduction by a forward bias voltage.

Integrated circuit IC2 operates as a voltage controlled resistor which forms an attenuator in conjunction with resistors R10 and R11. Resistor R10 limits the maximum attenuation to about 20 dB, which gives what most people consider to be the best tremolo effect. The output of the low frequency oscillator is connected to the input of integrated circuit IC2 by way of VR2. The latter is needed to attenuate the otherwise excessive output level from IC1. It is adjusted to give a reasonably smooth and symmetrical modulation characteristic, and this is just a matter of using trial and error to arrive at what is subjectively judged to be the best effect.

Integrated circuit IC3 is an output buffer stage, which also provides a small amount of voltage gain to compensate for the general losses through the voltage controlled amplifier. Switch S1 provides in/out switching, and it simply bypasses the unit altogether when the effect is switched out. The current consumption of the circuit is only about 3.5 mA from the 9 V PP3-sized battery.

Construction of the unit should not be difficult, but bear in mind that the 4007UBE used for integrated circuit IC2 is a CMOS device, and that it consequently requires the standard anti-static handling precautions to be observed. This device must be the unbuffered type having a *UBE* suffix. Ideally a unit of this type should be housed in a diecast aluminium box. A box of this type is very tough

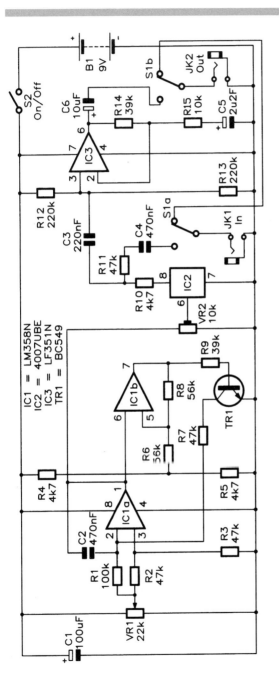

Figure 1.4 The alternative tremolo unit circuit diagram

IC1 = LM358N
IC2 = 4007UBE
IC3 = LF351N
TR1 = BC549

and provides excellent screening properties. A folded aluminium box will suffice, and is much cheaper. Switch S1 should be a heavy-duty pushbutton switch mounted on the top panel of the case so that it can be operated by foot. Most switches of this type are of the successive operation variety (i.e. operating the switch once switches out the effect, a second operation switches the effect back in again, a third operation switches it out once more, and so on).

Alternative tremolo unit parts list

Resistors — All 0.6 W 1% metal film (unless stated)

R1	100 k	1	(M100K)
R2,3, 7,11	47 k	4	(M47K)
R4,5,10	4k7	3	(M4K7)
R6,8	56 k	2	(M56K)
R9,14	39 k	2	(M39K)
R12,13	220 k	2	(M220K)
Rl5	10 k	1	(M10K)
VR1	22 k lin pot	1	(FW03D)
VR2	10 k min hor preset	1	(UH03D)

Capacitors

C1	100 µF 10 V axial electrolytic	1	(FB48C)
C2,4	470 nF polyester layer	2	(WW49D)
C3	220 nF polyester layer	1	(WW45Y)
C5	2µ2F 100 V PC electrolytic	1	(FF02C)
C6	10µF 50 V PC electrolytic	1	(FF04E)

Semiconductors

IC1	LM358N	1	(UJ34M)
IC2	HCF4007UBEY	1	(QX04E)
IC3	LF351N	1	(WQ30H)
TR1	BC549	1	(QQ15R)

Music projects

Miscellaneous

S1	press toe DPDT	1	(FH93B)
S2	SPST ultra-min toggle switch	1	(FH97F)
JK1,2	$^1/_4$ in jack skt brk	2	(IIF90X)
B1	9 V PP3-sized battery	1	(FK58N)
	battery clip	1	(HF28F)
	8-pin DIL socket	2	(BL17T)
	14-pin DIL IC socket	1	(BL18U)

Dynamic tremolo unit

A conventional tremolo unit is described earlier — this type of tremolo effect is still used a fair amount, but is considered rather *tame* by many. It is certainly a much more simple effect than, say, flanging or phasing. A more interesting tremolo effect can be obtained by varying the tremolo rate in some way. Manual variations can actually be achieved quite easily using an ordinary tremolo unit with the modulation rate potentiometer controlled via a pedal mechanism. Automatic variations can be provided by a very low frequency modulation oscillator or, as in this case, linking the tremolo rate to the envelope of the input signal. The higher the input level, the faster the tremolo rate. When used with a guitar this gives a high initial rate, but the rate steadily decreases as each note from the guitar decays. As with practically any effect which is linked to the dynamic level of the signal, this gives an interesting and quite *natural* effect.

It is not very difficult to add dynamic rate control to the tremolo circuit, due to the use of a voltage controlled oscillator (VCO) to provide the amplitude modulation signal. Instead of providing the control voltage from a potentiometer it merely has to be generated by a simple envelope follower circuit. Basically, this just means using a simple rectifier and smoothing circuit to convert the input signal to a corresponding d.c. voltage. This circuit is based on diodes Dl and D2. Germanium diodes are used in the rectifier circuit as they have a lower forward voltage drop than silicon diodes. This gives more accurate tracking at low signal levels. The attack time of

Music projects

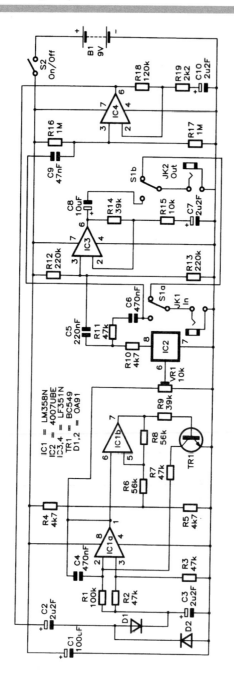

Figure 1.5 The dynamic tremolo unit circuit diagram

this circuit is very short, and although the decay time is much longer, it is still short enough to give accurate tracking on practically any input signal.

Quite a large voltage swing is needed in order to take the VCO over a reasonably wide frequency range. As a result of this the output level from most guitars is too low to drive the rectifier and smoothing circuit at an adequate level. Integrated circuit IC4 is therefore used ahead of the rectifier circuit to provide buffering and a boost signal level. The specified value for resistor R18 is suitable for low output guitar pick-ups, but for operation with high output pickups R18 will need to be much lower in value (around 10 k). With very low output pick-ups a higher value might be beneficial (about 220 k). The current consumption of this circuit is about 4.5 mA, a 9 V PP3-sized battery is therefore adequate as the power source.

When constructing the unit bear in mind that the 4007UBE used for IC2 is a CMOS device which consequently requires the standard anti-static handling precautions to be observed. Also remember that diodes D1 and D2 are germanium diodes which are vulnerable to heat damage. Extra care should therefore be taken when soldering these components into circuit.

Music projects

Dynamic tremolo unit parts list

Resistors — All 0.6 W 1% metal film (unless stated)

Rl	100 k	1	(M100K)
R2,3,7,11	47 k	4	(M47K)
R4,5,10	4k7	3	(M4K7)
R6,8	56 k	2	(M56K)
R9,14	39 k	2	(M39K)
R12,13	220 k	2	(M220K)
R15	10 k	1	(M10K)
R16,17	1 M	2	(M1M)
R18	120 k	1	(M120K)
R19	2k2	1	(M2K2)
VR1	10 k min hor preset	1	(UH03D)

Capacitors

C1	100 µF 10 V axial electrolytic	1	(FB48C)
C2,3,7,10	2µ2F 100 V PC electrolytic	4	(FF02C)
C4,6	470 nF poly layer	2	(WW49D)
C5	220 nF poly layer	1	(WW45Y)
C8	10 µF 50 V PC electrolytic	1	(FF04E)
C9	47 nF poly layer	1	(WW37S)

Semiconductors

IC1	LM358N	1	(UJ34M)
IC2	HCF4007UBEY	1	(QX04E)
IC3,4	LF351N	2	(WQ30H)
TR1	BC549	1	(QQI5R)
D1,2	OA91	2	(QH72P)

Miscellaneous

S1	press toe DPDT	1	(FH93B)
S2	SPST ultra-min toggle switch	1	(FH97F)
JK1,2	$^1/_4$ in jack skt brk	2	(HF90X)
B1	9 V PP3-sized battery	1	(FK58N)
	battery clip	1	(HF28F)
	8-pin DIL socket	3	(BL17T)
	14-pin DIL socket	1	(BL18U)

Music projects

Metal pedal

While not perhaps to everyone's taste, the so-called *metal* sound is now an established electronic musical effect. Technically it is produced by mixing two signals creating sum and difference frequencies, neither of which are normally harmonically related to the original input frequencies. The result can be very discordant, or with the two input frequencies spaced at suitable musical intervals the result is a very rich sound, like bells, gongs, and similar metallic instruments (hence the *metal* name).

The effect is simple in theory, but in practice it can be more difficult to get really good results. The usual set-up is to have a variable frequency oscillator feeding one input of a mixer (or *ring modulator* as it is generally known in this application), with the output from a guitar or other instrument feeding into the second input. The ring modulated signal is then mixed into the straight-through signal in the required quantity. The usual problem is that of breakthrough of the oscillator signal at the output, and even if the ring modulator gives a respectable 40 dB or so of carrier suppression this still leaves a clearly audible signal under no input conditions. A noise gate is therefore normally required in order to render the breakthrough inaudible.

In this design, the need for a noise gate is avoided by having a very high degree of carrier suppression in the ring modulator, and the latter is actually the phase comparator of an NE565N phase locked loop (IC2). The VCO stage of this device acts as the carrier oscillator, and it provides a frequency range of approximately 100 Hz to 2 kHz with RV1 acting as the frequency control. RV2 is

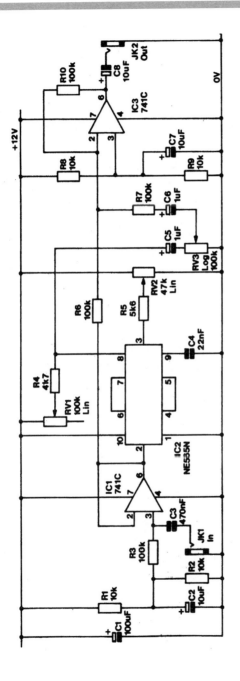

Figure 1.6 Metal pedal circuit diagram

Music projects

the carrier balance control, and this is carefully adjusted to minimise the carrier breakthrough at the output. It is a front panel control rather than a preset, as the optimum setting varies very slightly with changes in the carrier frequency, and although a good compromise setting can be found it was felt to be better to have the ability to null out the carrier as much as possible. The maximum degree of carrier suppression seems to be quite high, with around 80 dB being achievable with the prototype.

IC1 simply acts as an input buffer stage which provides the unit with an input impedance of about 100 kΩ. Its output is direct-coupled to the input of IC2, for which it provides with the necessary input bias voltage. IC3 acts as a conventional summing mode mixer circuit which combines the modulated and straight-through signals. RV3 controls the amount of ring modulated signal that is mixed into the unprocessed signal. Signal levels of up to about 8 volts peak-to-peak can be handled using a 12 volt supply (the use of a 9 volt battery supply is not recommended). Current consumption is about 10 milliamps.

Metal pedal parts list

Resistors — All 0.6 W 1% metal film

R1,2,8,9	10 k	4	(M10K)
R3,6, 7,10	100 k	4	(M100K)
R4	4k7	1	(M4K7)
R5	5k6	1	(M5K6)
RV1	100 k lin pot	1	(FW05F)
RV2	47 k lin pot	1	(FW04E)
RV3	100 k log pot	1	(FW25C)

Capacitors

C1	100 µF 35 V axial electrolytic	1	(FB49D)
C2,7,8	10 µF 50 V PC electrolytic	3	(FF04E)
C3	470 nF poly layer	1	(WW49D)
C4	22 nF poly layer	1	(WW33L)
C5,6	1 µF 100 V PC electrolytic	2	(FF01B)

Semiconductors

IC1,3	LM741CN (8-pin DIL)	2	(QL22Y)
IC2	LM565N	1	(WQ56L)

Miscellaneous

JK1,2	$1/4$ in jack skt open	2	(HF91Y)
	DIL socket 8-pin	2	(BL17T)
	DIL socket 14-pin	1	(BL18U)

Headphone amplifier

This simple amplifier is intended for guitar practice purposes. It has a low power output stage which is suitable for use with medium impedance headphones (the type widely sold as replacements for personal stereo units). These offer good efficiency, but mostly have a wide frequency response and quite low distortion. In fact they generally offer remarkable audio quality for their price. Their lightweight construction makes them easy on the head and ears in prolonged use. This amplifier can be built to suit high or low output guitar pickups. It incorporates a simple *fuzz* effect which generates *soft* clipping. This gives what is generally accepted as a more musical effect than the harsh hard clipping produced by many home constructor fuzz effect units.

Integrated circuit IC1 acts as the pre-amplifier, and this is a standard operational amplifier inverting mode circuit. The voltage gain is controlled by negative feedback network resistors R1–R4, and the specified values give a nominal voltage gain of 20 dB (ten times). This should give the unit adequate sensitivity for operation with even the lowest output pickups. However, if necessary the gain can be boosted slightly by making resistor R4 a little higher in value (say 1 M). Rather than a lack of gain, it is more likely that the unit will prove to be too sensitive if it is used with a modern pickup having a high output level. There is then a risk of integrated circuit IC1 being driven into clipping and producing severe distortion. In order to avoid this, if the unit is to be used with high output pickups the value of resistor R4 should be reduced to 47 k. Integrated circuit IC1 then merely acts as a unity gain buffer stage.

The output of IC1 is coupled to the volume control by capacitor C4. Most electric guitars have a built-in volume control, but in order to obtain optimum noise performance the guitar's volume control should be set at maximum and the volume should be controlled using potentiometer RV1. The unit does not include any tone controls, but this is not too important as they are also a feature of most electric guitars. The output amplifier is another operational amplifier circuit, but this time it uses the non-biFET device, but it has been selected for this circuit simply because it has output characteristics that seem well suited to driving medium impedance headphones. It gave good volume when tried with several pairs of medium impedance headphones, but some other operational amplifiers (e.g. the LF351) were less successful. I would not recommend the use of alternative operational amplifiers in the integrated circuit IC2 position of this circuit.

The *soft fuzz* effect is introduced by closing switch S1, so that diodes D1 and D2 are added into integrated circuit IC2's negative feedback circuit. This gives what is a conventional clipping amplifier, except that diodes D1 and D2 are germanium diodes and not silicon types. Whereas silicon diodes require a forward bias of about 0.6 to 0.65 volts before they will conduct, and thereafter they conduct very heavily, germanium diodes start to conduct at a much lower voltage. Also, they turn on much more gradually. This gives a good *soft* clipping effect and a good *fuzz* sound. When the fuzz effect is switched in, resistor R8 is shunted across resistor R7 so that integrated circuit IC2's voltage gain is boosted. This ensures that there is sufficient gain to give reasonably strong

Music projects

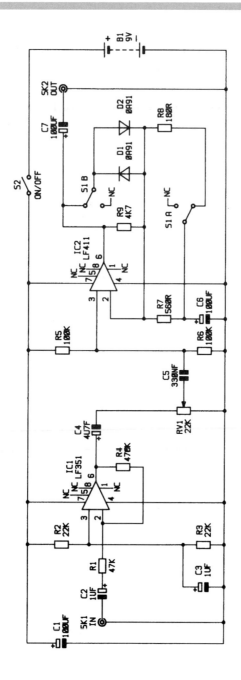

Figure 1.7 Headphone amplifier circuit diagram

clipping. It also compensates for the increased feedback introduced by the diodes, and the drop in volume it would otherwise introduce.

The headphones are driven from the output of integrated circuit IC2 via d.c. blocking capacitor C7, and results will probably be best if the phones are wired in series (i.e. ignore the earth tag and connect the other two tags to the output of the amplifier). The current consumption of the circuit is only about 5 milliamps under quiescent conditions, and two or three milliamps more than this at high volume levels. A PP3-sized 9 volt battery is adequate as the power source, and the unit can be built into a *pocket* size case if desired.

Music projects

Headphone amplifier parts list

Resistors — All 0.6 W 1% metal film

R1	47 k	1	(M47K)
R2,3	22 k	2	(M22K)
R4	470 k	1	(M470K)
R5,6	100 k	2	(M100K)
R7	560 Ω	1	(M560R)
R8	180 Ω	1	(M180R)
R9	4k7	1	(M4K7)
RV1	22 k log pot	1	(FW23A)

Capacitors

C1	10 µF 10 V axial electrolytic	1	(FB48C)
C2,3	1 µF 100 V PC electrolytic	2	(FF01B)
C4	4µ7F 63 V PC electrolytic	1	(FF03D)
C5	330 nF polyester	1	(WW47B)
C6,7	100 µF 10 V PC electrolytic	2	(FF10L)

Semiconductors

IC1	LF351N	1	(WQ30H)
IC2	LF411CN	1	(QY27E)
D1,2	OA91	2	(QH72P)

Miscellaneous

SK1	$^1/_4$ in jack skt open	1	(HF91Y)
SK2	3.5 mm stereo jack socket	1	(FK20W)
B1	9 V PP3-sized battery	1	(FK58N)
S1	DPDT ultra-min toggle	1	(FH99H)
S2	SPST ultra-min toggle	1	(FH97F)
	battery clip	1	(HF28F)
	8-pin DIL socket	2	(BL17T)

Music projects

Treble booster

Treble boost is one of the most simple of guitar effects, and as its name implies, it merely provides some high frequency boost to the processed signal. This gives a *brighter* sound, but without the distortion produced when a *fuzz* type unit is used. It is perfectly safe to play chords when using a treble booster! On the face of it you could get much the same effect by simply advancing the treble control on the amplifier. In practice this might result in treble boost being applied to other instruments as well, depending on the precise set up used. Anyway, the degree of treble boost available from an effects unit is substantially more than that which can be obtained with full treble boost on most amplifiers. It is perhaps worth pointing out that a unit of this type is reliant on there being some high frequency signals to boost. It can not *make a silk purse out of a sow's ear*, and if used with a really *muddy* pick-up it is unlikely to have much effect on the sound.

The circuit is basically just an operational amplifier used as a non-inverting amplifier. At low frequencies the circuit has approximately unity voltage gain due to the feed-back through resistor R4. At higher frequencies some of this feed-back is decoupled by resistor R1, potentiometer RV1, and capacitors C2 or C3. The low values of the capacitors results in the gain of the circuit rising steadily at 6 dB per octave through the treble range. The higher value of capacitor C3 results in it starting to apply the boost at a lower frequency than capacitor C2, thus giving a stronger effect with some middle frequencies being boosted significantly. The strength of the effect can also be controlled using potentiometer RV1, which limits the maximum amount of boost applied. It gives a boost range from a minimum of about 6 dB at maximum resistance, up to a little over 20 dB at minimum resistance. A higher maximum boost can be obtained by reducing the value of resistor R1. Bear in mind though, that using high levels of treble boost might give problems with feedback or excessive noise. Switch S2 enables the effect to be switched in and out.

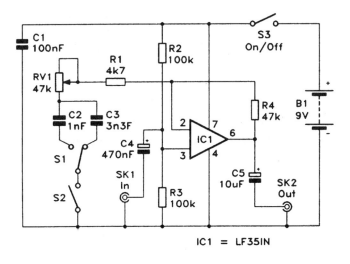

IC1 = LF35IN

Figure 1.8   The treble booster circuit diagram

Construction of this simple project is mostly straight-forward. However, bear in mind that when used with some guitar pick-ups it will be handling fairly low signal levels, and that the wiring must be kept quite short so that significant pick up of stray signals is avoided. A

Music projects

metal case earthed to the negative supply rail will provide screening and further assist in avoiding unwanted pick up. It is advisable to use a strong case as switch S2 should be a heavy duty push-button switch mounted on the top panel of the unit so that it can be operated by foot. The most convenient type of switch to use is the successive operation type. With this type of switch, operating it once switches the effect in, operating it again switches out the effect, a further operation switches it back in again, and so on. This avoids having to keep your foot on the switch for the duration that the effect is required. On the other hand, a simple push-to-make non-locking switch is best if you will need to repeatedly switch the effect in and out with precise timing.

The current consumption of the circuit is only about 2 milliamps, and a PP3-sized battery is therefore more than adequate as the power source. Some modern high output guitar pick-ups have surprisingly high output levels, and for operation with these the treble booster might be lacking in *headroom*. Any problems with overloading can be overcome by powering the unit from two 9 volt batteries wired in series so as to give an 18 volt supply.

Treble booster parts list

Resistors — All 0.6 W 1% metal film

R1	4k7	1	(M4K7)
R2,3	100 k	2	(M100K)
R4	47 k	1	(M47K)
RV1	47 k lin pot	1	(FW04E)

Capacitors

C1	100 nF ceramic disc	1	(BX03D)
C2	1 nF poly layer	1	(WW22Y)
C3	3n3F poly layer	1	(WW25C)
C4	470 nF 100 V PC electrolytic	1	(FF00A)
C5	10 µF 50 V PC electrolytic	1	(FF04E)

Semiconductor

IC1	LF351N	1	(WQ30H)

Miscellaneous

S1	SPDT sub-min toggle	1	(FH98G)
S2	press toe SPDT switch	1	(FH92A)
S3	SPST ultra-min toggle	1	(FH97F)
B1	9 V PP3-sized battery	1	(FK58N)
SK1,2	$1/_4$ in jack skt open	2	(HF91Y)
	battery clip	1	(HF28F)
	8-pin DIL socket	1	(BL17T)

Dynamic treble booster

Most treble boost effects units provide a preset amount of treble boost continuously. There is a potential problem with such units in that the treble boost can result in problems with feedback and noise, due to the greatly increased gain at some frequencies. The feedback can be combated by being careful with the positioning of loudspeakers relative to the instruments, and using the other normal methods of counteracting this problem.

The noise problem can be overcome by using a noise gate which can be a separate unit, or built into the effects unit, as in this dynamic treble booster. The noise will still be there when a reasonably strong signal level is present, but it will be masked by the main signal. When the signal decays to a low level, the signal path is cut and the noise is removed. In this case the unit does not provide a true noise gate action, and at low signal levels the signal path is maintained. However, the treble boost is removed on low level signals, which should reduce the noise to an acceptable level. This more subtle approach is not normally apparent, which is not always the case with a true noise gate action.

The basic treble booster circuit is exactly the same as the one featured previously.

However, the drain to source resistance of transistor TR1 has been added in series with switches S1, S2, etc., and the treble booster action is only obtained if transistor TR1 is switched on. This is a VMOS transistor, and accordingly it is normally switched off and requires a forward bias to bring it into conduction. Its control signal is produced by first amplifying the input signal using integrated circuit IC2, then rectifying and smoothing the amplified signal using diodes D1 and D2 in a conventional rectifier and smoothing circuit. This gives a control signal that is roughly proportional to the input signal level. This circuit has a fast attack time, plus a slower (but still reasonably fast) decay time, so that the d.c. output signal accurately tracks the input level. With no input signal or only a very low input level, the bias fed to transistor TR1's gate is too small to switch it on, and no treble boost is obtained. At more than quite a modest input level the bias on transistor TR1's gate is high enough to bias this device hard into conduction so that the full treble boost is obtained.

The notes on constructing the treble booster unit apply equally to this circuit. Additionally, note that diodes D1 and D2 are germanium diodes, and that they are more vulnerable to heat damage than are the more familiar silicon types. Consequently, extra care should be taken when soldering these components into circuit. Transistor TR1 is a MOS device, but it has a built in anti-static protection circuit.

In use it might be found that the value of resistor R6 has to be altered in order to give a suitable threshold level at which the treble boost is introduced. If the treble boost is held on by the background noise level, then resistor R6 must be made lower in value. If the treble boost is introduced only on volume peaks (which is unlikely), then the value of R6 must be increased. The current consumption of the circuit is only about 3 milliamps, and a

Music projects

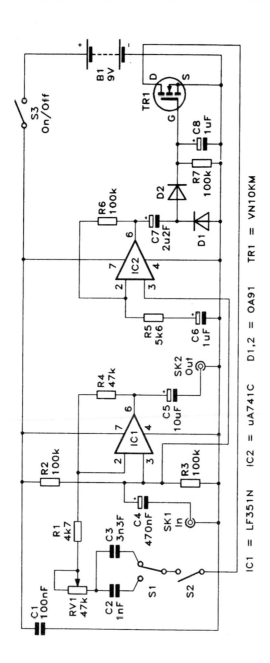

Figure 1.9   The dynamic treble booster circuit diagram

PP3-sized battery is adequate to power the unit. Like the basic treble boost unit, it might be necessary to use two 9 volt batteries in series to give an 18 volt supply if the unit is used with a high output guitar pick-up.

Music projects

Dynamic treble booster parts list

Resistors — All 0.6 W 1% metal film

R1	4k7	1	(M4K7)
R2,3,5,6	100 k	4	(M100K)
R4	47 k	1	(M47K)
R5	5k6	1	(M5K6)
RV1	47 k lin pot	1	(FW04E)

Capacitors

C1	100 nF ceramic disc	1	(BX03D)
C2	1 nF poly layer	1	(WW22Y)
C3	3n3F poly layer	1	(WW25C)
C4	470 nF 100 V PC electrolytic	1	(FF00A)
C5	10 µF 50 V PC electrolytic	1	(FF04E)
C6,8	1µF 100 V PC electrolytic	2	(FF01B)
C7	2µ2F 100 V PC electrolytic	1	(FF02C)

Semiconductors

IC1	LF351N	1	(WQ30H)
IC2	LM741CN	1	(QL22Y)
D1,2	OA91	2	(QH72P)
TR1	VN10KM	1	(QQ27E)

Miscellaneous

S1	SPDT sub-min toggle	1	(FH98G)
S2	press toe SPDT switch	1	(FH92A)
S3	SPST ultra-min toggle	1	(FH97F)
B1	9 V PP3-sized battery	1	(FK58N)
SK1,2	$^1/_4$ in jack skt open	2	(HF91Y)
	battery clip	1	(HF28F)
	8-pin DIL socket	2	(BL17T)

Compressor

A guitar compressor is an effects unit which merely compresses the dynamic range of the input signal, giving a virtually constant volume for the duration of each note. This gives a less *twangy* sound with a much longer sustain period. In fact units of this type are sometimes called *sustain* units, due to this elongated sustain period.

The circuit is based on a voltage controlled attenuator (VCA). The voltage controlled attenuator uses transconductance operational amplifier integrated circuit IC2 in a standard configuration. Strictly speaking this is a current controlled attenuator, as the gain is proportional to the control current fed to pin 1 of IC2. Resistor R11 biases the linearising diodes of IC2, and resistor R12 supplies a bias current to the control input. The bias via resistor R12 means that under standby conditions the voltage controlled attenuator provides about unity voltage gain.

The voltage controlled attenuator is preceded by a pre-amplifier stage which uses integrated circuit IC1 in the non-inverting mode. With the specified value for resistor R4 the circuit has a voltage gain of about 48 times, and this gives good results with my electric guitar. However, this is fitted with low output pick-ups, and with many guitars much lower voltage gain will be needed in order to avoid overloading and severe distortion at the output of preset variable resistor VR1. For high output pick-ups a value of around 22 k to 47 k would be more appropriate. With most pick-ups, the output level from integrated circuit IC2 will be somewhat higher than the input level. Preset VR1 can attenuate the output signal, and it is adjusted to give an output level that is roughly comparable to the input level. There is no need for any precise measurements here — it is just a matter of adjusting preset VR1 to give what is subjectively assessed to be the same volume with and without the unit connected in the signal path.

Some of the output from integrated circuit IC2 is fed to a non-inverting amplifier based on integrated circuit IC3. This is mainly required to act as a buffer amplifier, but it also provides a small amount of voltage gain (6 dB). The output from integrated circuit IC3 is coupled to a rectifier and smoothing network based on diodes D1 and D2. This produces a positive d.c. signal that is roughly proportional to the amplitude of the input signal. If this signal is strong enough, it biases transistor TR1 into conduction, which results in some of the control bias current being tapped off. This reduces the bias current fed to integrated circuit IC2, which in turn results in increased attenuation through integrated circuit IC2. This gives a simple feedback action, where an increased output level causes more current to be tapped off through transistor TR1, and the gain to be reduced. Once the input signal is high enough to bring transistor TR1 into conduction, this feedback action tends to hold the output signal at an almost constant level.

The value of resistor R15 controls the attack time of the circuit, and this has a strong influence on the effect obtained. A lower value tends to give over-shoot which totally suppresses the initial transient on each note. With the specified value the initial transients are allowed to pass, but at reduced volume, giving a conventional com-

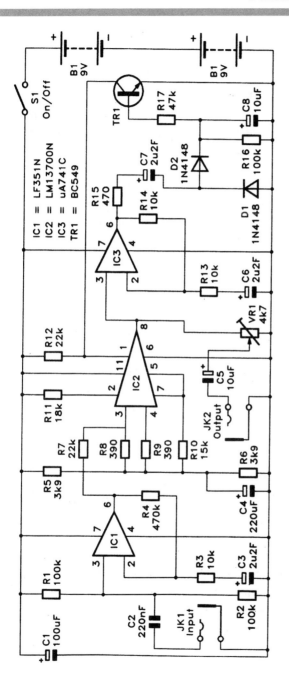

Figure 1.10 The compressor circuit diagram

49

pression effect. A higher value of around 680 or 820 ohms permits the transients to pass largely unaffected. This retains the *twangy* guitar sound to a large extent, but still gives the elongated sustain period. I personally prefer this effect, and it might be worthwhile experimenting with different values for resistor R15 to find the one which gives the effect you like best.

No method of switching out the effect is shown in the circuit diagram, but this is easily achieved. A conventional bypass switching arrangement using a double-pole, double-throw switch is probably the best method to use.

The circuit is powered from two 9 volt batteries connected in series to give an 18 volt supply. The unit *will* operate from a single 9 volt battery, but this gives less *headroom* at the output of integrated circuit IC1, and might result in severe distortion with some guitars. The current consumption using an 18 volt supply is about 8 milliamps.

In use, bear in mind that a unit such as this inevitably increases the overall gain in the system. This increases the risk of excessive *hum* pick-up and problems with feedback. Extra care therefore needs to be taken with the positioning of the speakers, guitars, mains leads, and so on.

Compressor parts list

Resistors — All 0.6 W 1% metal film

R1,2,16	100 k	3	(M100K)
R3,13,14	10 k	3	(M10K)
R4	470 k (see text)	1	(M470K)
R5,6	3k9	2	(M3K9)
R7,12	22 k	2	(M22K)
R8,9	390 Ω	2	(M390R)
R10	15 k	1	(M15K)
R11	18 k	1	(M18K)
R15	470 Ω (see text)	1	(M470R)
Rl7	47 k	l	(M47K)
VRl	4k7 hor encl preset	l	(UH02C)

Capacitors

Cl	100 μF 35 V axial electrolytic	1	(FB49D)
C2	220 nF poly layer	1	(WW45Y)
C3,6,7	2μ2F 100 V PC electrolytic	3	(FF02C)
C4	220 μF 35 V PC electrolytic	1	(JL22Y)
C5,8	10 μF 50 V PC electrolytic	2	(FF04E)

Semiconductors

ICl	LF35lN	l	(WQ30H)
IC2	LMl3700N	l	(YH64U)
IC3	LM741CN	l	(QL22Y)
TRl	BC549	l	(QQ15R)
Dl,2	1N4148	2	(QL80B)

Music projects

Miscellaneous

Sl	SPST ultra min toggle	1	(FH97F)
JK1,2	$^{1}/_{4}$ in jack skt open	2	(HF9lY)
Bl,2	9 V PP3-sized battery	2	(FK58N)
	8-pin DIL IC socket	2	(BLl7T)
	14-pin DIL IC socket	1	(BL18U)

Wah-wah effects unit

A wah-wah effects unit (or waa-waa if you prefer) is a simple form of filter effect. It is produced using a bandpass filter which is swept up and down the audio range, giving a sort of nasal wah-wah sound. Traditionally a wah-wah unit is controlled via a pedal which permits manual control of the filter's centre frequency.

With units that are designed for home construction an alternative method of control is often used. This means having some form of automatic filter control, such as sweeping its centre frequency via an envelope follower or a low frequency oscillator. These methods do not necessarily give better results, and are probably inferior to having manual control by a talented player. However, they do avoid having to build the pedal mechanism, which is an awkward aspect of construction unless you are very good at the mechanical side of project construction.

This wah-wah circuit is for a traditional pedal-type effects unit, which means that it is very simple from the electronic point of view. It is more difficult as far as the mechanical aspects of construction are concerned, but a ready-built pedal unit, such as that recommended makes the job easy. The recommended unit — as supplied — is fitted with a 100 k potentiometer and a screened cable fitted with a jack plug; these should be removed. A 1 k linear potentiometer should be fitted in place of the original one.

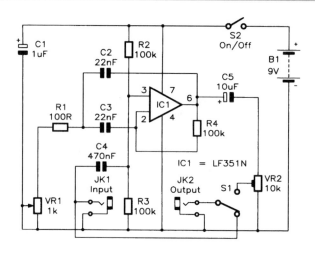

Figure 1.11 The wah-wah effects unit circuit diagram

The circuit is very straightforward, and the filter is based on a well known configuration. In fact this filter is usually encountered in its inverting form, but for the present application the non-inverting mode is better. This permits a fairly high (50 k) and constant input impedance to be achieved without having to resort to an input buffer stage. Potentiometer VR1 controls the filter's centre frequency, and permits it to be swept over the middle and upper-middle part of the audio spectrum. Resistor R4 is large in comparison with the resistance through VR1 and R4 so as to give the filter a suitably high Q value. This results in a significant voltage gain through the circuit. Potentiometer VR2 is adjusted to reduce the output level so that there is no significant change in volume when

54

switching the effect in or out. In/out switching is provided by switch S1, and this simply switches the output socket between the output signal from potentiometer VR2 and the direct signal from the guitar.

Current consumption of the circuit is only about one milliamp or so, and each PP3-sized battery will therefore have an extremely long operating life. If the unit is used with high output guitar pick-ups there is a risk of the unit becoming overloaded, which results in strong distortion on signal peaks. Powering the unit from two PP3 batteries wired in series provides an 18 V supply, and much more *headroom*. Alternatively, simply backing off the volume control of the guitar slightly should cure the problem.

Music projects

Wah-wah unit parts list

Resistors — All 0.6 W 1% metal film

R1	100 Ω	1	(M100R)
R2,3,4	100 k	3	(M100K)
VR1	1 k lin pot	1	(FW00A)
VR2	10 k min hor preset	1	(UH03D)

Capacitors

Cl	1 µF 100 V axial electrolytic	1	(FB12N)
C2,3	22 nF poly layer	2	(WW33L)
C4	470 nF poly layer	1	(WW49D)
C5	10 µF 50 V PC electrolytic	1	(FF04E)

Semiconductor

IC1	LF351N	1	(WQ30H)

Miscellaneous

S1	press toe SPST switch	1	(FH92A)
S2	SPST ultra-min toggle switch	1	(FH97F)
SK1,2	$^1/_4$ in jack skt brk	2	(HF90X)
B1	9 V PP3-sized battery	1	(FK58N)
	battery clip	1	(HF28F)
	8-pin DIL IC socket	1	(BL17T)
	pedal mechanism	1	(XY28F)

Bass fuzz

Ordinary *fuzz* or *distortion* effects units will work with bass guitars, or with an ordinary electric guitar when played at the lower end of its compass, but results are often not very good. This is simply due to the high frequency harmonics generated by the clipping of the signal tending to dominate and mask the low frequency fundamental signal, giving a sound that is often just a harsh buzzing noise. This can sometimes be used to good effect, but a somewhat less harsh effect is probably of greater all round use. There are two possible approaches to the problem, one of which is to use a so called *soft* clipping circuit which generates relatively weak high frequency harmonics. The alternative, and the approach adopted in this design, is to use an ordinary *hard* clipping circuit followed by a lowpass filter to remove the unwanted harmonics. Although the unit was primarily designed with bass guitars in mind, it will also work well with ordinary electric guitars giving an interesting variation on the standard distortion effect.

The circuit is very straightforward and breaks down into two sections; the clipping circuit based on integrated circuit IC1, and the lowpass filter built around integrated circuit IC2. Taking the clipping circuit first, this is an operational amplifier connected in the non-inverting mode. The circuit uses a single 9 volt supply with resistors R1, R2 and capacitor C1 providing a 4.5 volt tapping for biasing purposes. Resistor R4 biases the non-inverting input of IC1 and sets the input impedance at 100 k. Potentiometer RV1 and resistor R3 are the negative feedback network, and RV1 enables the voltage gain to be varied from unity at minimum value to around 48 times at maximum resistance. However, if the output signal exceeds about 1.2 volts peak-to-peak, diodes D1 and D2 are brought into conduction on signal peaks (D1 on negative peaks, D2 on positive peaks). When brought into conduction the diodes effectively shunt RV1, reducing the feedback resistance and voltage gain of the circuit, and introducing the required distortion. In practice RV1 is adjusted to give a level of gain which ensures that the input signal drives the circuit into clipping.

The lowpass filter is a high slope type, and is actually a fourth order (24 dB per octave) circuit. In other words, above the cut-off frequency, a doubling of frequency results in the gain reducing by a factor of sixteen. The cut-off frequency is at a little over 2 kHz. This gives a good effect with the middle frequency harmonics left unaffected, but the high frequency harmonics virtually

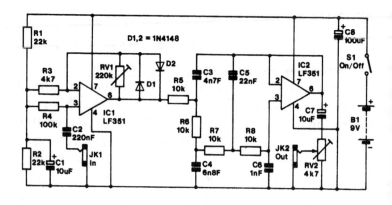

Figure 1.12   Bass fuzz circuit diagram

eliminated. However, if desired the cut-off frequency can be changed by altering the value of resistors R5 to R8. Changes in value give an inversely proportional change in the cut-off frequency (e.g. doubling the resistor's value to 20 k would reduce the cut-off frequency to just over 1 kHz).

Potentiometer RV2 is adjusted to give an output level which is comparable to the direct output of the guitar, although no bypass switch is shown in the circuit diagram, a standard double-pole double-throw footswitch bypass circuit can be added if required. Potentiometer RV1 is given any setting which provides adequate gain to give a well-clipped signal, but does not make the setup over sensitive with any slight vibration of the guitar producing strong unwanted output signals. Noise can sometimes be a problem with fuzz units, but in this case, the use of low noise devices and heavy lowpass filtering at the output results in excellent noise performance.

Bass fuzz parts list

Resistors — All 0.6 W 1% metal film

R1,2	22 k	2	(M22K)
R3	4k7	1	(M4K7)
R4	100 k	1	(M100K)
R5–8	10 k	4	(M10K)
RV1	220 k hor encl preset	1	(UH07H)
RV2	4k7 hor encl preset	1	(UH02C)

Capacitors

C1,7	10 µF 50 V PC electrolytic	2	(FF04E)
C2	220 nF poly layer	1	(WW45Y)
C3	4n7F poly layer	1	(WW26D)
C4	6n8F poly layer	1	(WW39N)
C5	22 nF poly layer	1	(WW33L)
C6	1 nF poly layer	1	(WW22Y)
C8	100 µF 10 V PC electrolytic	1	(FF10L)

Semiconductors

IC1,2	LF351N	2	(WQ30H)
D1,2	1N4148	2	(QL80B)

Miscellaneous

JK1,2	$^1/_4$ in jack skt brk	2	(HF90X)
S1	SPST ultra min toggle	1	(FH97F)
B1	9 V PP3-sized battery	1	(FK58N)
	battery clip	1	(HF28F)
	8-pin DIL socket	2	(BL17T)

2 General effects

Music projects

Lowpass filter effects unit

There are a great many forms of musical effects units, and most rely on some form of frequency selective filtering for their operation. The unit featured here is in this category, and it is basically just a 12 dB per octave lowpass filter which is swept by a low frequency oscillator. This gives a sort of tremolo effect on the high frequency content of the processed signal, producing a relatively mild but useful effect. It is an effect that is available on many synthesisers, but which seems to be something of a rarity as far as stand-alone effects units are concerned.

The circuit breaks down into two sections; the filter which is built around integrated circuit IC1, and the oscillator which is based on integrated circuits IC2 and IC3. Starting with the filter, this uses two transconductance operational amplifiers which are contained in a single LM13700N device. A Darlington pair emitter follower output stage is also included for each amplifier, and these have discrete load resistors R8 and R12.

The two amplifiers are connected in series, and in this application function more as voltage controlled resistors than amplifiers. They act as simple 6 dB per octave lowpass filters in conjunction with capacitors C3 and C4, giving a combined attenuation rate of 12 dB per octave. Feedback through resistors R6 and R7 gives what is actually a bandpass response at pin 8 of integrated circuit IC1, and by taking the output signal from here a form of wah-wah effect can be obtained. Integrated circuit IC1 is current rather than voltage operated, but the inclusion of resistor R14 in series with the control inputs gives a current flow that is roughly proportional to the applied voltage, and effectively converts the filter to voltage controlled operation. Resistor R13 reduces the input voltage range from the oscillator slightly, bringing it into a more suitable range to drive the filter.

The oscillator uses integrated circuit IC3 in a well known configuration which is based on a Miller integrator IC3(a) and a Schmitt trigger IC3(b). This type of circuit gives both squarewave and triangular outputs. In this application a triangular waveform gives good results with a smooth sweeping of the filter frequency, and it is this output that is utilised. The operating frequency can be varied by means of potentiometer RV2, and the nominal frequency range is from 10 Hz at minimum resistance to 0.2 Hz (one cycle every five seconds) at maximum resistance.

It is more than a little useful to have some control over the sweep range, and this is provided by potentiometer RV1. This controls the feedback applied to integrated circuit IC2, and hence the voltage gain of this device. When set at a low value IC2 has only a low voltage gain, resulting in the cut-off frequency of the filter being varied over a narrow range of frequencies in the lower treble range. Higher resistance gives greater sweep width, with the cut-off frequency being swept over most of the audio frequency range with potentiometer RV1 set at maximum value.

As with any effects unit, it is advisable to build the unit into a strong metal case such as a diecast aluminium type. If a bypass switch is needed a standard double-

Music projects

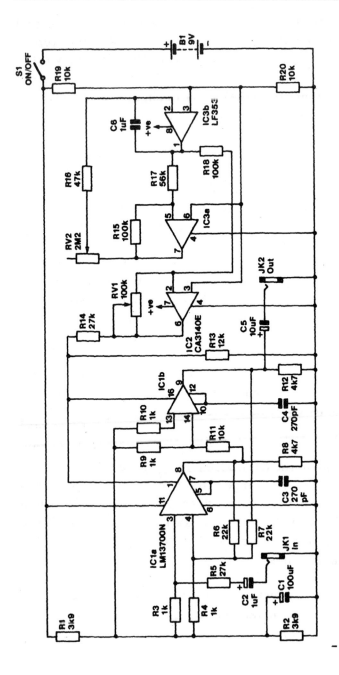

Figure 2.1 Low pass filter diagram

pole, double-throw bypass configuration can be used, and the switch should be a heavy duty push button type mounted on the top of the case so that it can be operated by foot.

Music projects

Low pass filter parts list

Resistors — All 0.6 W 1% metal film

R1,2	3k9	2	(M3K9)
R3,4,9,10	1 k	4	(M1K0)
R5,14	27 k	2	(M27K)
R6,7	22 k	2	(M22K)
R8,12	4k7	2	(M4K7)
R11,19,20	10 k	3	(M10K)
R13	12 k	1	(M12K)
R15,18	100 k	2	(M100K)
R16	47 k	1	(M47K)
R17	56 k	1	(M56K)
RV1	100 k lin pot	1	(FW05F)
RV2	2M2 lin pot	1	(FW09K)

Capacitors

C1	100 µF 10 V axial elect	1	(FB48C)
C2	1 µF 100 V PC elect	1	(FF01B)
C3,4	270 pF ceramic	2	(WX61R)
C5	10 µF 25 V axial elect	1	(FB22Y)
C6	1 µF poly layer	1	(WW53H)

Semiconductors

IC1	LM13700N	1	(YH64U)
IC2	CA3140E	1	(QH29G)
IC3	LF353	1	(WQ31J)

Miscellaneous

S1	SPST min toggle	1	(FH97F)
JK1,2	$1/4$ in jack skt open	2	(HF91Y)
B1	9 V PP3-sized battery	1	(FK58N)

Accented metronome

The conventional pendulum type metronome now seems to be something of a dying breed, it has been superseded by electronic devices that generate the regular train of *click* sounds. Some electronic instruments now have a built-in metronome, and it is quite easy to produce a simple stand-alone unit for use with instruments that lack this facility. The design featured here has a frequency range of approximately 0.5 to 5 Hertz, or about 30 to 300 beats per minute in other words. Some mechanical metronomes have the ability to emphasise every *ninth* beat, usually by ringing a bell on the accentuated beats. This unit has a similar feature, but it produces a low pitched *thud* sound on the accented beats instead of the usual *click* sound. Every second, third, or fourth beat can be stressed, or this feature can be switched out altogether if desired. The unit can easily be modified to accentuate anything from every second beat to every ninth beat if required.

A low frequency oscillator generates the procession of *click* sounds, and this oscillator is a simple 555 astable type based on integrated circuit IC1. Potentiometer RV1 is the frequency (beat rate) control. Resistor R2 has been made very low in value so that the output signal from integrated circuit IC1 is a series of very brief pulses. This gives the required high pitched *click* sound. The loudspeaker is driven from the output via an emitter follower buffer stage (transistor TR2). Integrated circuit IC1 provides short negative output pulses, but what we require here is positive pulses. This is nothing to do with the sound produced, which is the required *clicks* in either case. It is a matter of ensuring that the current to the loudspeaker is switched off most of the time, and that it is only driven during the brief output pulses. This gives a low current consumption, whereas the alternative of having the loudspeaker activated for most of the time would give a massive current consumption. Transistor TR1 acts as a simple inverter to provide the output stages with pulses of the correct polarity.

The accentuation is obtained by feeding the output pulses from IC1 to a divide by *N* circuit. This is based on integrated circuit IC2 which is a decade counter and one-of-ten decoder. It is made to divide by two, three, or four by feeding the appropriate one-of-ten output to the re-set input. The required division rate is set using switch S1. If no accentuation is required, S1 is set to the *0* posi-tion. The reset input is then connected to the *0* output, which holds the counter permanently in the reset state. When the accentuation is active the output pulses from output *0* are shaped by capacitor C3 together with di-odes D1 and D2, and mixed with the output pulses from integrated circuit IC1. Their longer pulse duration gives them a lower pitch than the ordinary output pulses, and they are also reproduced at a slightly higher volume level which helps to make them stand out still further.

The current consumption of the circuit is about 9 milliamps, which is mainly the current drawn by IC1. A small (PP3-sized) battery is adequate as the power source, but if the unit is likely to receive a great deal of use, a higher capacity type would give lower running costs.

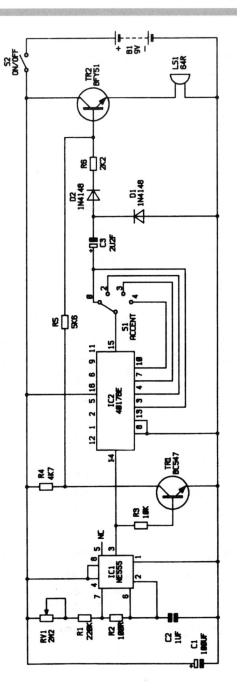

Figure 2.2 Accented metronome circuit diagram

Music projects

Construction of the unit does not provide any major difficulties, but bear in mind that integrated circuit IC2 is a CMOS device, and that it consequently requires the standard anti-static handling precautions to be observed. Potentiometer RV1 should be fitted with a large control knob so that it can be equipped with a calibrated scale of reasonable accuracy. Finding the calibration points is a matter of counting the number of beats in a given period of time in order to determine the beat rate, plus a certain amount of trial and error in order to get RV1 precisely set to the desired calibration rates.

As pointed out previously, you can obtain accentuation on any beat from every second one to every ninth beat. It is just a matter of using a switch having the required number of ways and using the appropriate outputs of integrated circuit IC2. Table 2.1 shows which output pins provide which division rates. Whether you consider such rates as seven and nine worthwhile is your own decision!

IC2 pin number	Division rate
4	2
7	3
10	4
1	5
5	6
6	7
9	8
11	9

Table 2.1   Division rates

Metronome parts list

Resistors — All 0.6 W 1% metal film

R1	220 k	1	(M220K)
R2	100 Ω	1	(M100R)
R3	10 k	1	(M10K)
R4	4k7	1	(M4K7)
R5	5k6	1	(M5K6)
R6	2k2	1	(M2K2)
RV1	2M2 lin pot	1	(FW09K)

Capacitors

C1	100 µF 10 V axial electrolytic	1	(FB48C)
C2	1 µF poly layer	1	(WW53H)
C3	2µ2F 100 V axial electrolytic	1	(FB15R)

Semiconductors

IC1	NE555N	1	(QH66W)
IC2	HCF4017BEY	1	(QX09K)
TR1	BC547	1	(QQ14Q)
TR2	BFY51	1	(QF28F)
D1,2	1N4148	2	(QL80B)

Miscellaneous

S1	4-way 3-pole switch	1	(FH44X)
S2	SPST ultra-min toggle	1	(FH97F)
LS1	66 mm dia 64 ohm speaker	1	(WF57M)
B1	9 V PP3-sized battery	1	(FK58N)
	battery clip	1	(HF28F)
	8-pin DIL IC holder	1	(BL17T)
	16-pin DIL IC holder	1	(BL19V)

Dual tracking effects unit

In the past Mullard's TDA1022 and TDA1097 *bucket brigade* delay-line integrated circuits have been popular for use in effects units and other circuits that require audio signals to be delayed by around 1 to 100 milliseconds. Unfortunately, these two chips were taken out of production a few years ago, and are no longer generally available. In fact they are probably unobtainable from anywhere now. However, the MN3XXX series of delay-line chips are still available, and would seem able to handle virtually anything that was in the repertoire of the two Mullard delay-line chips. The delay-line device used in this project is an MN3207, which is a 1024 stage type.

This unit provides a dual tracking effect, or with a few extra parts (see the following project), a chorus effect. In essence it is a very simple effect that takes the input from a single instrument and provides an output that sounds like two instruments playing in unison. It is an effect that can be used to enrich virtually any signal, but it gives the most obvious improvement on fairly simple sounds. For example, it works well with single VCO synthesiser sounds, electric guitars, or even a human voice (via a suitable microphone and pre-amplifier). In fact it is probably most used with solo voice signals to produce an instant duet. The effect is produced by mixing the input signal with a delayed version of itself.

Integrated circuit IC3 is the delay-line chip, while IC6 is the matching clock and bias generator chip. The clock frequency determines the delay time and the maximum bandwidth that can be used. In this case a frequency of around 25 kHz has been selected. The delay time is equal to the number of delay-line stages divided by double the clock frequency. This gives a delay time of around 20 milliseconds, which is suitable for the dual tracking effect. A delay of at least 10 milliseconds is required in order to produce the doubling-up effect, but the effect must be no longer than 60 milliseconds as this gives a short echo effect instead.

The clock frequency must be at least double the maximum input frequency, but should preferably be three or more times the maximum input frequency. Input signals at excessive input frequencies cause a severe form of distortion known as *aliasing* distortion. A 25 kHz clock rate means that the audio bandwidth of the circuit must be limited to around 8 kHz for good results. This is substantially less than the full audio bandwidth, but is wide enough to give good results. It should be borne in mind here that it is only the delayed signal that has this restricted bandwidth. The straight-through signal has the full audio bandwidth, which results in the unit as a whole having no obvious lack of high-frequency response.

The delay-line must be preceded by a low-pass filter which prevents high frequency signals from causing problems. This filter is a standard three stage (18 dB per octave) type based around integrated circuit IC2. A buffer stage using integrated circuit IC1 is included ahead of the filter. This provides the unit with an input impedance of about 50 kΩ, and ensures that the filter is fed from a suitably low source impedance. Potentiometer VR1 controls the biasing of IC1, and most of the rest of the circuit for that matter, as it is largely d.c. coupled.

Music projects

VR1 is adjusted to give optimum large signal handling, and with a 9 volt supply voltage swings of about 3 to 4 volts peak-to-peak can be accommodated.

A low-pass filter is also needed at the output of the unit. Resistors R8 and R9 mix the output signals from the last two stages of integrated circuit IC3, and this helps to minimise the clock breakthrough at the output. However, it is an inevitable consequence of the sampling delay-line technique that the output signal is stepped. The low-pass filtering smooths out this stepping to provide an ordinary audio output signal. The filter is a four stage (24 dB per octave) active type based around integrated circuit IC4. Integrated circuit IC5 operates as a conventional summing mode mixer which combines the delayed and non-delayed signals. Switch S1 can be used to switch out the effect by blocking the delayed signal from the mixer. The current consumption of the circuit is approximately 9 milliamps. A PP3-sized battery is just about adequate as the power source, but a higher capacity type such as six HP7-sized cells in a plastic battery holder would be a better option if the unit is likely to receive a lot of use.

When constructing the unit do remember that integrated circuits IC3 and IC6 are both MOS devices, which therefore require the normal anti-static handling precautions. If the unit is built as a pedal unit, switch S1 should be a heavy-duty push-button switch mounted on the top panel of the case, so that it can be operated by foot. The case should be a tough type, such as a diecast aluminium box. If suitable test equipment is available, preset VR1 should be set for symmetrical clipping using the normal techniques. In the absence of suitable test gear, simply give

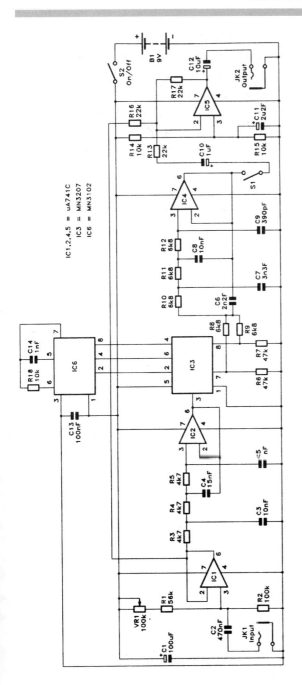

Figure 2.3 The dual tracking effects unit circuit diagram

Music projects

preset VR1 any setting that gives low distortion on high level signals. Note that the unit should be fed with high level signals of around one volt r.m.s. if it is to achieve a good signal-to-noise ratio (pre-amplified signals). With signal sources such as microphones and low output guitar pick-ups the unit must be preceded by an appropriate pre-amplifier.

Dual tracking unit parts list

Resistors — All 0.6 W 1% metal film

R1	56 k	1	(M56K)
R2	100 k	1	(M100K)
R3,4,5	4k7	3	(M4K7)
R6,7	47 k	2	(M47K)
R8,9,10, 11,12	6k8	5	(M6K8)
R13,16, 17	22 k	3	(M22K)
R14,15, 18	10 k	3	(M10K)
VR1	100 k min hor preset	1	(UH06G)

Capacitors

C1	100 µF 10 V axial electrolytic	1	(FB48C)
C2	470 nF poly layer	1	(WW49D)
C3,8	10 nF poly layer	2	(WW29G)
C4	15 nF poly layer	1	(WW31J)
C5,14	1 nF poly layer	2	(WW22Y)
C6	2n2F poly layer	1	(WW24B)
C7	3n3F poly layer	1	(WW25C)
C9	390 pF polystyrene	1	(BX52G)
C10	1 µF 100 V PC electrolytic	1	(FF01B)
C11	2µ2F 100 V PC electrolytic	1	(FF02C)
C12	10µF 50 V PC electrolytic	1	(FF04E)
C13	100 nF 16 V ceramic disc	1	(YR75S)

Music projects

Semiconductors

IC1,2,4,5	LMA741CN	4	(QL22Y)
IC3	MN3207	1	(UR67X)
IC6	MN3102	1	(UR68Y)

Miscellaneous

JK1,2	$^1/_4$ in jack skt open	2	(HF91Y)
S1	SPST ultra min toggle	1	(FH97F)
S2	press toe SPDT switch	1	(FH92A)
B1	9 V PP3-sized battery	1	(FK58N)
	battery clip	1	(HF28F)
	8-Pin DIL IC socket	6	(BL17T)

Chorus unit

This project is basically the same as the dual tracking unit just described. The only difference between these two effects is that the delay time of a dual tracking unit is fixed, whereas the delay time used in a chorus unit is varied at a low frequency. The difference between the two effects can be subtle, or very obvious. This depends on both the modulation frequency used, and the signal source. By varying the delay time, the effect obtained is an output that sounds like numerous voices or instruments playing in unison, rather than just the two of the dual tracking effect.

Something that should not be overlooked is that the variations in the delay time produces a certain amount of vibrato. In other words, there is some frequency modulation on the output signal. The practical importance of this is that with a polyphonic input signal the musical intervals between the input frequencies are altered slightly. This can give some slightly discordant results. Unless you like that type of thing, it is best to either switch out the modulation when processing polyphonic signals, or to use fairly low modulation frequencies.

The circuit of the chorus unit is the same as that of the dual tracking unit, but the low frequency oscillator shown here must be added. This oscillator uses a standard configuration that is based on a Miller integrator IC7(a) and a trigger circuit IC7(b). There is a squarewave output from IC7(b), and a triangular output from IC7(a). In this application the triangular signal is the more suitable signal. It is coupled to the clock oscillator via switch S3 and resistor R23.

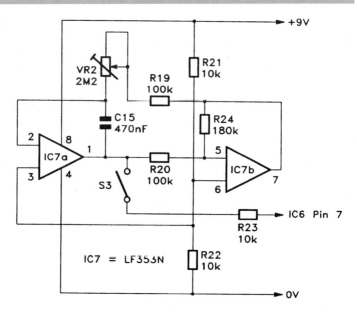

Figure 2.4 The additional circuitry required to make the dual tracking effects unit produce a chorus effect

Switch S3 enables the modulation to be switched in and out, while resistor R23 controls the modulation depth. The value of resistor R23 has been made quite low so that reasonably strong modulation is produced. A higher value will give reduced modulation, but would probably give too little modulation to be worthwhile. A lower value might give stronger modulation, but there is a risk that it would cause the clock oscillator to cut off for a portion of each half cycle. Although the clock oscillator chip does not actually have provision for voltage control, loosely coupling a modulation signal to its C–R timing network gives the desired effect. Connecting the modulation signal to pin 7 of the clock chip seems to give the best modulation range.

Variable resistor VR2 enables the operating frequency of the modulation oscillator to be adjusted from about 0.5 Hz at maximum resistance, to about 10 Hz at minimum resistance. The effect tends to be much stronger at the high frequency end of the range than at low modulation frequencies. When initially testing the unit it is therefore a good idea to set VR2 at minimum resistance, so that the strongest (and most obvious) effect is obtained. A PP3-sized battery is suitable for the power supply.

Music projects

Chorus unit parts list

Resistors — All 0.6 W 1% metal film

R19,20	100 k	2	(M100K)
R21,22,23	10 k	3	(M10K)
R24	180 k	1	(M180K)
VR2	2M2 lin pot	1	(FW09K)

Capacitors

C15	470 nF poly layer	1	(WW49D)

Semiconductors

IC7	LF353N	1	(WQ31J)

Miscellaneous

S3	SPST ultra min toggle	1	(FH97F)

In addition to the above, a full set of the dual tracking unit components detailed previously are required.

3 Synthesisers

Featuring:

Music projects

Woodblock synthesiser

Woodblock synthesisers are perhaps not the best known type of percussion synthesiser, but they are capable of producing useful effects and are something electronic music enthusiasts should not overlook. This woodblock synthesiser can be triggered either by hitting the pick-up or by means of a +5 volt trigger pulse of a few milliseconds in duration. The pitch is adjustable over a wide frequency range.

A woodblock sound requires a fairly complex waveform because it consists of not just a fundamental signal plus harmonics, but a fundamental signal plus other frequencies close to this fundamental one. A suitable waveform cannot be produced using an oscillator, and the most simple method of generating a satisfactory signal is to use a noise source plus a very narrow bandwidth bandpass filter.

In this circuit the basic noise signal is generated by the reverse biased base/emitter junction of transistor TR1. This breaks down at about 6 volts in zener diode fashion, and like a zener diode generates noise spikes. The noise output is greater than that from a zener diode though, but the high level of amplification provided by common emitter transistor amplifier TR2 is still needed in order to give a strong output. Transistor TR1 can be virtually any npn silicon transistor, and in practice it is probably best to try a few devices from the spares box to find one that gives a really good noise output. Capacitor C3 provides some initial lowpass filtering that prevents the final output from having an excessive high frequency content.

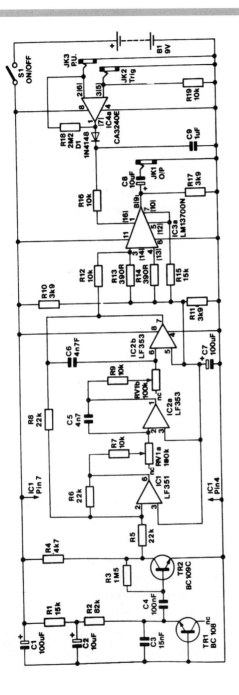

Figure 3.1 Woodblock synthesiser circuit diagram

Music projects

The output from transistor TR2 is fed to what is really a state variable filter (and which is based on integrated circuits IC1 and IC2). However, in this case it is only the lowpass output of the filter that is utilised — the highpass and bandpass outputs just being ignored. The filter is used at a high Q value so that a narrow and pronounced peak is produced in the frequency response just below the cut-off frequency, and the type of filtering obtained is actually a cross between bandpass and lowpass filtering. This gives an excellent output signal for this application, and the pitch of the output can be varied from about 200 Hz to 20 kHz by means of potentiometer RV1.

Envelope shaping is, of course, essential if a usable effect is to be obtained. In this circuit IC3(a) operates as the voltage controlled amplifier, and this is a straightforward circuit using an operational transconductance amplifier. Integrated circuit IC4 and its associated components generate a simple attack — decay control voltage from the input trigger pulse or the pulses from the pickup. The latter can be a ceramic resonator or a crystal microphone insert.

One section of IC3 is left unused, and the pin numbers for the unused section are shown in brackets on the circuit diagram. In practice the unit would be most useful if two or more synthesisers are constructed, and the second section of IC3 does not have to be wasted. The noise source can be used for several synthesisers, but the filter, envelope generator, and voltage controlled amplifier circuits would need to be duplicated for each synthesiser. If several synthesisers are built and they

86

are to be triggered by pick-ups it is essential to have these reasonably well (physically) isolated from one another to prevent unwanted multiple triggering from occurring.

Woodblock synthesiser parts list

Resistors — All 0.6 W 1% metal film

R1,15	15 k	2	(M15K)
R2	82 k	1	(M82K)
R3	1M5	1	(M1M5)
R4	4k7	1	(M4K7)
R5,6,8	22 k	3	(M22K)
R7,9,12,			
16,19	10 k	5	(M10K)
R10,11,			
17	3k9	3	(M3K9)
R13,14	390 Ω	2	(M390R)
R18	2M2	1	(M2M2)
RV1	100 k dual pot lin	1	(FW88V)

Capacitors

C1,7	100 µF 10 V axial electrolytic	2	(FB48C)
C2,8	10 µF 25 V axial electrolytic	2	(FB22Y)
C3	15 nF polyester	1	(BX71N)
C4	100 nF polyester	1	(BX76H)
C5,6	4n7F poly layer	2	(WW26D)
C9	1 µF poly layer	1	(WW53H)

Semiconductors

IC1	LF351N	1	(WQ30H)
IC2	LF353N	1	(WQ31J)
IC3	LM13700N	1	(YH64U)
IC4	CA3240E	1	(WQ21X)
TR1	BC108C	1	(QB32K)
TR2	BC109C	1	(QB33L)
D1	1N4148	1	(QL80B)

Miscellaneous

JK1,2,3	$^{1}/_{4}$ in jack skt open	3	(HF91Y)
S1	SPST ultra min toggle	1	(FH97F)
B1	9 V PP3-sized battery	1	(FK58N)
	battery clip	1	(HF28F)

Music projects

Modulated syndrum

The original *Syntom* and two companion syndrum projects are old projects, originally published in *Electronics — The Maplin Magazine* several years ago now. These have proved to be most popular projects, which is not really surprising when you consider their low cost and the wide range of sounds they can generate. These projects produce various fixed or swept pitch sounds, such as disco-drum, wave, and gong sounds. Although they are very versatile, there are some types of sound which they cannot produce. In particular, the pitch of sounds can only be at a fixed frequency, or swept in sympathy with the envelope of the output signal. There is no provision for modulated sounds generated with the aid of a low frequency oscillator (LFO).

This syndrum circuit is loosely based on the original *Syntom* circuit, but instead of having a voltage controlled oscillator (VCO) controlled from the envelope generator circuit, it is fed from the triangular output of an LFO. This permits a useful range of *warbling* sounds to be produced. Like the original design the unit is touch sensitive and it is activated by tapping the case.

The basic tone is generated by a simple VCO based on integrated circuit IC2. This is similar to the standard triangular/squarewave oscillator configuration, with IC2(a) acting as the integrator and IC2(b) operating as the trigger circuit. However, an additional transistor (TR1) is used, and together with some extra resistors at the input this provides voltage control. The control voltage is applied to the junction of resistors R6 and R7. It is the triangular output signal from IC2(a) that is used in this case, although the squarewave signal from IC2(b) could be used if preferred. The lower harmonic content of the triangular signal gives better sounds for most purposes.

A standard triangular/squarewave oscillator based on integrated circuit IC1 produces the low frequency modulation signal. IC1(a) acts as the integrator while ICl(b) operates as the trigger circuit. The triangular output signal from IC1(a) gives a good modulation characteristic. Again, the squarewave output could be used if preferred, but it would simply switch the VCO between two pitches, whereas the triangular signal provides a better, swept, effect. Potentiometer VR1 enables the output voltage range to be varied. When VR1 is well backed-off the VCO operates at low frequencies with a modest amount of modulation. Advancing VR1 gives a generally higher pitched output together with a greater modulation depth. Potentiometer VR2 enables the modulation frequency to be varied from under 1 Hz to around 10 Hz.

Integrated circuit IC3 together with transistor TR2 act as the voltage controlled attenuator (VCA), and these are used in a conventional transconductance amplifier configuration. The CA3080E used for IC3 has no built-in output buffer stage, making it necessary to include transistor TR2 as an emitter follower output stage. The control signal for the voltage controlled attenuator is produced by first using IC4(b) to amplify and half wave rectify the output from the microphone MIC1 (which is actually a piezoelectric sounder). A smoothing circuit then produces a proportional d.c. signal from the output of IC4(b). Potentiometer VR3 controls the decay time of this signal, which can be anything from a few hundred

Music projects

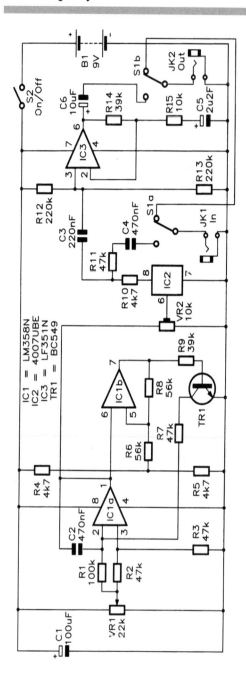

Figure 3.2   Modulated syndrum circuit diagram

milliseconds to around five seconds. IC4(a) simply acts as a buffer stage between the smoothing circuit and the control input of the voltage controlled attenuator.

The current consumption of the circuit is about 7 mA, and a PP3-sized battery is therefore adequate as the power source. In the prototype a cased ceramic resonator was used for microphone MIC1, but any resonator fixed on the inside of the case should work properly. With some resonators, the circuit might be oversensitive, but this can be corrected by making resistor R21 higher in value. The output from the unit is approximately 5 V peak-to-peak, which should be sufficient to drive any mixer or, power amplifier.

Music projects

Syndrum parts list

Resistors — All 1% 0.6 W metal film

R1,3,6, 19,20	100 k	5	(M100K)
R2,7,8,9	47 k	4	(M47K)
R4,5	3k3	2	(M3K3)
R10,11	56 k	2	(M56K)
R12	39 k	1	(M39K)
R13	27 k	1	(M27K)
R14,15	220 Ω	2	(M220R)
R16	10 k	1	(M10K)
R17	8k2	1	(M8K2)
R18	4k7	1	(M4K7)
R21	1 k	1	(M1K)
VR1	10 k lin pot	1	(FW02C)
VR2	I M lin pot	1	(FW08J)
VR3	2M2 lin pot	1	(FW09K)

Capacitors

C1	100 μF 10 V axial electrolytic	1	(FB48C)
C2	220 nF poly layer	1	(WW45Y)
C3	470 μF 16 V PC electrolytic	1	(FF15R)
C4	1 nF poly layer	1	(WW22Y)
C5	10 μF 50 V PC electrolytic	1	(FF04E)
C6	I μF 100 V PC electrolytic	1	(FF0lB)

Semiconductors

IC1	MC1458CN	1	(QH46A)
IC2,4	LM358N	2	(UJ34M)
IC3	CA3080E	1	(YH58N)
TR1,2	BC549	1	(QQ15R)
D1	1N4148	1	(QL80B)

Miscellaneous

MIC1	miniature piezo sounder	1	(FM59P)
S1	SPST ultra-min toggle switch	1	(FH97F)
JK1	$^1/_4$ in jack skt brk	1	(HF90X)
B1	9 V PP3-sized battery	1	(FK58N)
	battery clip	1	(HF28F)
	8-pin DIL socket	4	(BL17T)

Music projects

Syndrum interface

This circuit can be used in conjunction with a mono-phonic synthesiser having gate and CV inputs to effectively form a sophisticated syndrum capable of pro-ducing a wide range of interesting effects. The main function of the circuit is to generate a trigger pulse when the pick up is activated. The latter is a crystal micro-phone insert or a 27 mm piezoelectric transducer which is mounted on a drum, pad of rubber, or whatever. When operated, the first negative output half cycle from the transducer triggers the 555 monostable circuit based on integrated circuit IC2, and this gives a positive 5 volt trigger pulse of about 100 ms in duration at jack socket JK2 (a nominal 15 volt pulse can be obtained from pin 3 of IC2). This pulse length should be satisfactory, but practically any desired figure can be obtained by alter-ing the value of R5 and/or C5. The circuit can be triggered using a positive trigger pulse of between about 5 and 15 volts applied to JK1, and this permits manual and auto-matic triggering to be employed together.

With some synthesisers it may be possible to obtain ris-ing and falling pitch syndrum sounds without the use of an external control voltage circuit, but in many cases this would be difficult or impossible. The unit therefore incorporates a rising/falling control voltage generator. Integrated circuit operational amplifier IC3(a) amplifies the trigger signal to produce a strong positive output pulse that charges C7 to a potential of several volts. The exact charge potential depends on how hard the trans-ducer is struck, and the unit is to a degree touch sensitive in this respect. However, if switch S2 is set to the other

96

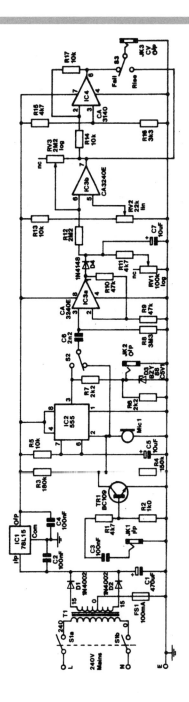

Figure 3.3   Syndrum interface circuit diagram

position IC3(a) is fed with the output of operational amplifier IC2, and capacitor C7 then charges to about 10 volts or so each time the unit is operated. Diode D3 prevents capacitor C7 from discharging into the output stage of IC3(a), and the decay time of the voltage is therefore largely controlled by potentiometer RV1. Operational amplifier IC3(b) is a level shifter and inverting amplifier, but it really acts as an attenuator as its voltage gain is never more than unity. Potentiometer RV2 acts as the pitch control while RV3 is used to control the sweep width. Operational amplifier IC4 is a straightforward unity gain inverter, and this enables a rising or falling voltage (pitch) to be selected using switch S3.

The circuit is powered from a simple 15 volt stabilised mains power supply unit which can comfortably provide the supply current of only about 20 to 25 milliamps.

The syndrum interface parts list

Resistors — All 0.6 W 1% metal film

R1,11,			
15	4k7	3	(M4K7)
R2	1k	1	(M1K)
R3	180 k	1	(M180K)
R4	150 k	1	(M150K)
R5,13,			
14,17	10 k	4	(M10K)
R6,7	2k2	2	(M2K2)
R8	3M3	1	(M3M3)
R9,10	47 k	2	(M47K)
R12	2M2	1	(M2M2)
R16	3k3	1	(M3K3)
RV1	100 k pot log	1	(FW25C)
RV2	22 k pot lin	1	(FW03D)
RV3	2M2 pot log	1	(FW29G)

Capacitors

C1	470 µF 35 V PC electrolytic	1	(FF16S)
C2,3,4	100 nF polyester	3	(BX76H)
C5,7	10 µF 50 V PC electrolytic	2	(FF04E)
C6	2n2F poly layer	1	(WW24B)

Music projects

Semiconductors

D1,2	1N4002	2	(QL74R)
D3	BZY88C5V1	1	(QH07H)
D4	1N4148	1	(QL80B)
TR1	BC109C	1	(QB33L)
IC1	LM78L15ACZ	1	(QL27E)
IC2	NE555N	1	(QH66W)
IC3	CA3240E	1	(WQ21X)
IC4	CA3140E	1	(QH29G)

Miscellaneous

T1	15 V 6VA min transformer	1	(WB15R)
FS1	100 mA fuse 20 mm	1	(WR00A)
	fuseholder 20 mm	1	(RX96E)
S1	switch rocker DPST	1	(YR69A)
S2,3	switch sub-min toggle A	2	(FH00A)
JK1–3	$1/_4$ in jack skt brk	3	(HF90X)
Mic 1	crystal earpiece	1	(LB25C)
	$1/_4$ in jack plug	3	(HF85G)

4 MIDI projects

Featuring:

MIDI through box

The standard method of driving pieces of MIDI equipment from a control device is the so-called *chain* system of connection. This has the *out* socket of the controller connected to the *in* socket on one of the other pieces of equipment, and then the *thru* socket of this device connects to the *in* socket of the next piece of equipment, and so on. In theory, any number of instruments can be connected together by wiring the *thru* socket of one instrument to the *in* socket of the next in the *chain*. In practice this is not always possible though. There can be problems with what are often called *delays*, but which are more probably problems with smearing of the signal that compromise reliability. At a more basic level, many items of MIDI equipment (especially keyboard instruments) simply do not have a MIDI *thru* port, and cannot be used with this method of connection. Actually, if only one instrument lacks a *thru* socket it is possible to use the *chain* system, provided this instrument is placed at the end of the *chain*. If more than one instrument lacks a *thru* port, then the *star* system must be used.

It is only possible to adopt the *star* system if the MIDI controller has multiple outputs, or a *thru* box is included in the system. This method of connection relies on the controlling device having an *out* socket for each MIDI input that must be driven. A *thru* box simply takes the signal from a MIDI output and splits it to give a number of *thru* outputs to drive the inputs of the other equipment in the system. A *thru* box cannot be a passive device as a MIDI output only provides a drive current of 5 milliamps, and splitting this between several inputs would give an insufficient drive current for each one.

In this circuit, an opto-isolator is used at the input. This is not strictly necessary as there is no need for a *thru* box to provide isolation, but MIDI outputs are designed to drive an opto-isolator, and this method ensures reliable operation with any output that properly meets the MIDI standard. The 6N139 used in the integrated circuit IC1 position is not a simple LED/transistor type, but on the output side actually has a photo-diode, an emitter follower transistor, and a common emitter output stage. This gives high efficiency and fast operating speed. The circuit can comfortably accommodate MIDI's fairly high baud rate of 31250 baud. Resistor R4 ensures that the emitter follower stage operates at a reasonable current and that the device achieves a suitably fast switching speed. The 5 milliamp drive current is set partly by resistor R1, and partly by a series resistor in the drive circuit.

On the output side of the circuit there are four common emitter switching transistors, with each one driving a separate *thru* socket. Two current limiting resistors are used in each output circuit, and this two resistor system gives better protection to the circuit in the event of a system being incorrectly wired up, or a fault occurring. Four output stages are shown in the circuit diagram, but integrated circuit IC1 is capable of driving several more output stages if necessary. Power is provided by a 6 volt battery (such as four HP7-sized cells in a plastic holder). The quiescent current consumption will probably be

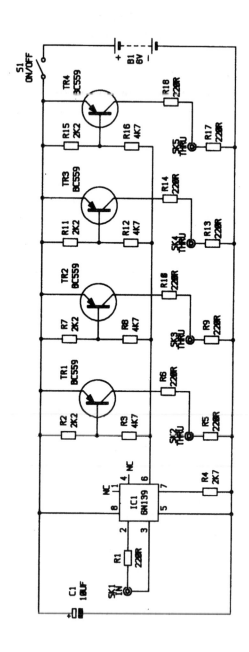

Figure 4.1 MIDI through box circuit diagram

negligible, but under worst-case conditions the average current drain could be as much as 2.5 milliamps per output that is actually used.

The standard MIDI connectors are 5-way (180 degree) DIN plugs and sockets. Provided you use the appropriate type of socket connected in the manner outlined, the *thru* box can be wired into the system using standard MIDI leads. If you make up your own leads, twin screened cable is required. Pins 2, 4, and 5 on one plug are connected to the corresponding pins of the other plug, with the screen carrying the connection between the two pin 2s. Note that some audio 5-way DIN leads use cross coupling and are unsuitable for MIDI applications.

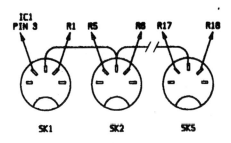

Figure 4.2   Socket connections

Music projects

MIDI through box parts list

Resistors — All 0.6 W 1% metal film

R1,5,6,9, 10,13,14, 17,18	220 Ω	9	(M220R)
R2,7,11, 15	2k2	4	(M2K2)
R3,8, 12,16	4k7	4	(M4K7)
R4	2k7	1	(M2K7)

Capacitor

C1	10 µF 25 V axial electrolytic	1	(FB22Y)

Semiconductors

IC1	6N139 opto isolator	1	(RA59P)
TR1,2, 3,4	BC559	4	(QQ18U)

Miscellaneous

SK1,2,3, 4,5	5-way (180°) DIN socket	5	(HH34M)
S1	SPST ultra-min toggle	1	(FH97F)
B1	1.5 V AA-sized battery	4	(FK55K)
	8-pin DIL socket	1	(BL17T)
	4 x AA battery holder	1	(HF29G)
	battery clip	1	(HF28F)

MIDI lead checker

Anyone who is involved with a MIDI system is likely to encounter problems before too long, with a unit in the system not responding to the signals sent to it. There are several possible causes for the lack of activity. The two most common ones are something in the system not being set up correctly, and a damaged MIDI lead. Checking for a broken lead is probably the best place to start, especially if several units in a *chained* system fail to operate. The likely cause of the problem is then a damaged lead feeding into the first unit in the *chain* which is failing to respond properly. The simplest of continuity testers are good enough for lead checking purposes, but investigating leads fitted with DIN plugs tends to be a very fiddly task.

MIDI lead testers are now available as ready-made products, but these are rather expensive. A simple do-it-yourself MIDI lead tester can be produced at quite low cost though, and is a more practical proposition for amateur MIDI users. Assuming the tester featured here is housed in a low cost box, or simply left as an open unit, it can be built for just a few pounds. It will indicate a lack of continuity between pairs of pins that should be interconnected, as well as showing up any short circuits between pins of a plug.

Basically all the unit has to do is feed a d.c. signal down each conductor in turn, with some form of indicator showing whether or not the signal is reaching the far end of the cable. In this circuit a five way switch is used to enable the signal to be manually connected to each conductor, one at a time. Resistor R1 provides current

Music projects

limiting for the five LED indicators at the opposite end of the cable. If there is continuity between the selected pin and a pin at the opposite end of the cable, then the appropriate LED will switch on. If there is a short circuit between the selected pin and another pin or pins, then two or more LEDs will switch on. If a pin on one plug connects to the wrong pin on the other plug, then this will be shown by the wrong LED lighting up. Of course, with a lack of continuity through a connector, no LEDs will be activated.

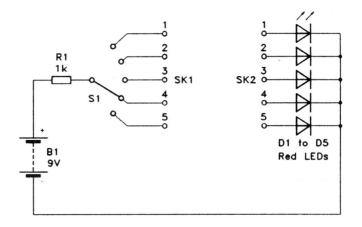

Figure 4.3   MIDI lead checker circuit diagram

With a normal MIDI cable there is a connection from pin 1 to pin 1, pin 2 to 2, and pin 3 to pin 3. Pins 4 and 5, which are the ones at the ends of the arc of pins, should not be connected. In reality it is not uncommon for these pins to be linked, but this should not prevent the lead from operating. These pins are simply not connected on pieces of MIDI equipment. Presumably a lot of MIDI leads are actually audio types and not leads made specifically for MIDI use. Some 5-way DIN audio leads have pins 1 to 5 connected to pins 5 to 1 respectively. These cross-connected leads are not suitable for MIDI use. Atari ST computers have a non-standard *thru/out* socket using all the pins of the 5-way DIN socket. With an ST *thru/out* lead pins 4 and 5 on the main plug connect to pins 1 and 3 on the *thru* plug. Note that a lack of connection between pin 2 on one plug and pin 2 on the other should not result in a MIDI lead failing to work. It does mean that its shielding is faulty though, and in use it might radiate electrical interference.

Construction should present few difficulties. The prototype is constructed on stripboard using printed circuit mounting sockets, but obviously the unit can be built using panel mounted sockets and LEDs, with everything hard-wired if preferred.

MIDI lead checker parts list

Resistors — All 0.6 W 1% metal film

| R1 | 1 k | 1 | (M1K) |

Semiconductors

| D1,2,3, 4,5 | Red LED | 5 | (WL27E) |

Miscellaneous

SK1,2	5-way 180° DIN skt	2	(YX91Y)
S1	12-way 1 pole rotary	1	(FF73Q)
B1	9 V PP3-sized battery	1	(FK58N)
	battery clip	1	(HF28F)

5 Amplifiers and pre-amplifiers

Featuring:

Music projects

Portable stereo amplifier

This amplifier is battery-powered and is built into a case about 500 mm wide which also houses the two loudspeakers. It is intended for use with a personal stereo cassette player, radio, or radio/cassette unit, and it enables loudspeaker operation to be obtained without losing portability. In effect it converts a personal stereo unit into a conventional stereo radio or cassette unit, but when maximum portability is needed the personal stereo unit alone can be used.

The circuit is based on an integrated circuit LM377N dual audio power amplifier device, and the two power amplifiers in this device are rather like operational amplifiers having high current output stages. The amplifiers are used in the non-inverting mode with the non-inverting outputs biased to an internal potential divider circuit of the LM377N which has its output at pin 1. Capacitor C9 provides decoupling of this bias voltage. The negative feedback loops give the amplifiers a voltage gain of only about 15 dB (5.5 times), and a higher level of voltage gain would be pointless because personal stereo units provide a fairly high output voltage of about 1 volt r.m.s. Using a 12 volt supply (eight 1.5 volt dry cells or preferably ten AA NiCad cells) the circuit provides a maximum output power of about 1 watt r.m.s. or so (per channel) at low distortion. The volume, balance, and tone are adjusted using the controls on the personal stereo unit.

When switch S2 is open resistor R7 results in the two output signals being mixed together to a certain extent, but they are mixed in such a way that signals in both channels tend to partially cancel out one another, and this boosts the channel separation. With the limited loudspeaker separation of the unit this can produce a better stereo effect, but how well or otherwise this system works depends to a large extent on the programme source used. However, in most cases it seems to give a more spacious and realistic effect.

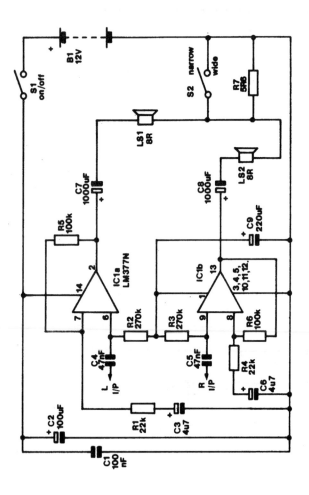

Figure 5.1 The portable stereo amplifier circuit diagram

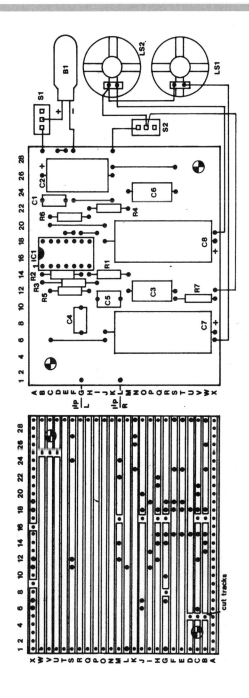

Figure 5.2 The breadboard layout and wiring for the portable stereo amplifier

115

The portable amplifier matrixboard assembly parts list

Resistors — All 0.6 W 1% metal film

R1,4	22 k	2	(M22K)
R2,.3	270 k	2	(M270K)
R5,6	100 k	2	(M100K)
R7	5Ω6	1	(M5R6)

Capacitors

C1	100 nF poly layer	1	(WW41U)
C2	100 µF 35 V axial electrolytic	1	(FB49D)
C3,6	4µ7F 100 V axial electrolytic	2	(FB18U)
C4,5	47 nF poly layer	2	(WW37S)
C7,8	1000 µF 16 V axial electrolytic	2	(FB82D)
C9	220 µF 16 V axial electrolytic	1	(FB61R)

Semi conductors

IC1	LM377N-9	1	(QH38R)

Miscellaneous

S1,2	SPDT sub-min toggle	2	(FH00A)
LS1,2	loudspeaker 8 Ω impedance	2	
B1	12 V battery 0.1 inch matrixboard plastic or metal case connection wire as required.		(BL09K)

Amplifiers and pre-amplifiers

Parametric equaliser

A parametric equaliser is a versatile form of tone control used principally in production of electronic music, but circuits of this type can also be used in hi-fi systems. Both lift and cut can be provided like an ordinary bass or treble tone control, but it is a frequency band somewhere in the middle of the audio range that is controlled by this type of filter, rather than one end of the audio spectrum. The centre frequency is tuneable (usually over a fairly wide frequency range), and the filter is really a bandpass and notch type, with the type of filtering provided depending on whether the circuit is set for lift or cut. Circuits of this type invariably have variable Q, so that a very narrow range of frequencies, a broad frequency range, or anything in between these two extremes can be controlled.

In electronic music, a parametric equaliser can obviously be used to radically alter the sound of an instrument, and can modify the sound in a variety of ways. When used with a hi-fi system it could be used to counteract a resonance or other irregularity in the frequency response of the system.

Although quite simple, the design featured here has a respectable level of performance with a tuning range which extends from about 200 Hz to approximately 4 kHz. Up to about 15 dB of boost and cut can be provided and the Q can be varied over wide limits.

In common with other designs of this general type, the circuit is based on a state variable filter. This is formed by IC1(a), IC2(a) and IC2(b), and it is the bandpass out-

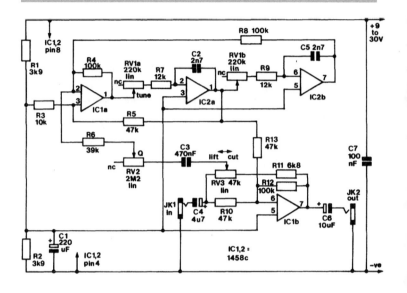

Figure 5.3   Parametric equaliser circuit diagram

put at pin 1 of IC2(a) that is utilised here. The frequency of the filter is governed by the values of C2 and C5, plus the series resistances of R7 plus RV1(a) and R9 plus RV1(b). By making the resistive elements variable, the operating frequency can be adjusted over the nominal range specified above, with minimum resistance corresponding to maximum operating frequency.

The signal is not actually handled directly by the bandpass filter, but instead passes through inverting amplifier IC1(b). The bandpass filter is effectively used as a sort of frequency selective network in the negative feedback circuit of IC1(b). The point of doing this is that it enables the notch response to be obtained in addition to the bandpass response, with the type of filtering obtained depending on whether RV3 is adjusted for lift or cut. Feedback over the filter (and hence its Q value) is controlled by R5, and R6 plus RV2. The Q can therefore be controlled using RV2, with minimum resistance corresponding to maximum Q (and a narrow response).

The circuit will operate on any supply voltage in the range 9 to 30 volts, but a supply of around 15 to 30 volts is preferable as it enables high output levels to be handled without clipping and serious distortion resulting. Bear in mind that when set for maximum boost the circuit provides a significant amount of voltage gain at the centre of the response, and it is then more vulnerable to overloading.

Music projects

Parametric equaliser parts list

Resistors — All 0.6 W 1% metal film

R1,2	3k9	2	(M3K9)
R3	10 k	1	(M10K)
R4,8,12	100 k	3	(M100K)
R5,10,13	47 k	3	(M47K)
R6	39 k	1	(M39K)
R7,9	12 k	2	(M12K)
R11	6k8	1	(M6K8)
RV1	220 k dual pot lin	1	(FW89W)
RV2	2M2 pot lin	1	(FW09K)
RV3	47 k pot lin	1	(FW04E)

Capacitors

C1	220 µF 63 V PC electrolytic	1	(FF14Q)
C2,5	2n7F 1% polystyrene	2	(BX61R)
C3	470 nF poly layer	1	(BX80B)
C4	4µ7F 63 V PC electrolytic	1	(FF03D)
C6	10 µF 50 V PC electrolytic	1	(FF04E)
C7	100 nF poly layer	1	(BX76H)

Semiconductors

IC1,2	MC1458CN	2	(QH46A)

Miscellaneous

JK1,2	¼ in jack skt brk	2	(HF90X)
	¼ in jack plug	2	(HF85G)

Microphone pre-amplifier

A perennial problem when using audio equipment is that of a component in the system which provides an inadequate output level to drive the input with which you would like to use it. One of the most common offenders is the humble microphone and all common types have an output level of a few millivolts r.m.s. or less. In fact, low impedance dynamic microphones, and some other types, have typical output levels of only a few hundred microvolts r.m.s. Many amplifiers and other items of audio equipment only have high level inputs which require a few hundred millivolts r.m.s., and (possibly) an RIAA equalised cartridge input which may have adequate sensitivity but is unusable anyway, due to the strong bass boost and treble cut of the equalisation.

The problem is easily overcome by using a suitable pre-amplifier to boost the signal to an adequate level to drive a high level input. This pre-amplifier circuit is designed for use with a low impedance (200 Ω to 1 k) dynamic microphone or a type which has similar characteristics (some electret types for instance). It provides a voltage gain of up to about 80 dB (10,000 times) and with a maximum output level of over 2 volts r.m.s. from a low source impedance, it can provide sufficient output to drive any normal high level input. Although the unit is inexpensive to build, it achieves a fairly high standard of performance with a good signal-to-noise ratio of around 70 dB under typical operating conditions.

A three stage circuit is used with transistors TR1 and TR2 acting as a low noise input stage and voltage amplifier. These are connected in a well known direct coupled

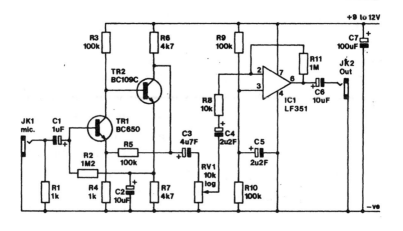

Figure 5.4 Microphone pre-amplifier circuit diagram

configuration, which has both devices in the common emitter mode. Transistor TR1 is operated at a low collector current of approximately 50 microamps in order to give a good signal-to-noise ratio. The noise level is not as low as can be obtained using one of the best audio operational amplifiers or pre-amplifier integrated circuits, but the noise performance is superior to that obtained using inexpensive operational amplifiers. However, the cost is comparable to inexpensive operational amplifiers and is far less than that of special low noise integrated circuits. Resistor R5 introduces negative feedback which reduces the voltage gain of the amplifier to about 40 dB (100 times) and gives improved distortion performance.

Capacitor C3 couples the output from transistor TR2 to gain control RV1, and from here the signal is coupled to the output amplifier. This is a straightforward inverting amplifier which, like the input stages, has a nominal voltage gain of 40 dB. The BIFET operational amplifier specified for integrated circuit IC1 gives good noise and distortion performance.

The circuit has a current consumption of about 3 to 4 milliamps, and a 9 volt battery is suitable as the power source. The input and output of the amplifier are out-of-phase, and problems with instability due to stray feedback are unlikely. It is still essential to keep all the wiring near the input of the unit as short as possible and to use a screened lead to connect socket JK1 to the circuit board. It is definitely advisable to house the unit in a case of all metal construction and earthed to the negative supply rail to provide overall screening against mains hum and other sources of electrical interference.

Microphone pre-amplifier parts list

Resistors — All 0.6 W 1% metal film

R1,4	1 K	2	(M1K)
R2	1M2	1	(M1M2)
R3,5,9,10	100 k	4	(M100K)
R6,7	4k7	2	(M4K7)
R8	10 k	1	(M10K)
R11	1M	1	(M1M)
RV1	10 k log pot	1	(FW22Y)

Capacitors

C1	1 µF 100 V PC electrolytic	1	(FF01B)
C2,6	10 µF 50 V PC electrolytic	2	(FF04E)
C3	4µ7F 63 V PC electrolytic	1	(FF03D)
C4,5	2µ2F 100 V PC electrolytic	2	(FF02C)
C7	100 µF 10 V PC electrolytic	1	(FF10L)

Semiconductors

IC1	LF351N	1	(WQ30H)
TR1	BC650	1	
TR2	BC109C	1	(QB33L)

Miscellaneous

JK1,2	3.5 mm mono jack skt	2	(HF82D)

Microphone pre-amplifier/limiter

Although this microphone pre-amplifier and limiter is very simple it provides a useful level of performance. It is primarily intended for use ahead of a mixer and tape deck, but it could, no doubt, be used in other, similar applications. Many tape decks have a built-in limiter, but there is a drawback in using a single limiter with several signal sources in that an overload from one source can effectively reduce the signal level provided by the other sources. Using a separate limiter for each input signal totally eliminates this problem.

The pre-amplifier is quite straight forward and uses integrated circuit IC1 as a non-inverting amplifier having a voltage gain of about 22 times, and integrated circuit IC2 as an inverting amplifier having a voltage gain of approximately 26 times. The circuit is intended for use with a high impedance microphone, and the total voltage gain is sufficient to give an output in excess of 1 volt r.m.s. with most microphones of this type. Potentiometer RV1 can be used to give a lower level of gain if necessary. An excellent signal-to-noise ratio is obtained by using a low noise bipolar operational amplifier in the IC1 position and a bi-FET type in the IC2 position.

An LM13700N dual transconductance operational amplifier is used as the basis of the limiter. IC3(a) is used as a straightforward current controlled amplifier and under quiescent conditions this is biased by resistor R15 so that it has approximately unity voltage gain. Resistor R14 biases the linearising diodes of IC3(a) and gives improved distortion and large signal handling performance.

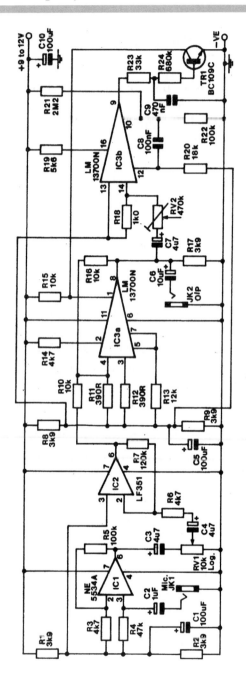

Figure 5.5 The microphone pre-amplifier/limiter circuit diagram

Amplifiers and pre-amplifiers

IC3(b) is used as a straightforward amplifier, and the gain of this stage can be varied by means of potentiometer RV2. Capacitor C8 couples the output of this amplifier to the input of the Darlington pair emitter follower output stage of IC3(b). This drives transistor TR1 via resistors R23 and R24, but under quiescent conditions transistor TR1 is cut off due to the low bias level supplied to the emitter follower buffer stage by resistors R21 and R22.

However, on positive output half cycles from IC3(b), provided the signal level is high enough, transistor TR1 will be biased into conduction. It then reduces the bias current to IC3(a) so that its gain is reduced and the required limiting action is obtained. Capacitor C9 integrates the pulses from IC3(b), but the values of capacitor C9, and resistors R23 and R24 have been chosen to give fast attack and decay times so that the limiting action is unlikely to be noticed unless a very severe overload occurs.

Potentiometer RV2 is adjusted to give the desired limiting level, and this can be any practical value above about 20 millivolts r.m.s. Raising the input signal 20 dB above the limiting threshold gives an increase of less than 2 dB at the output. A substantial overload could result in the signal at the output of integrated circuit IC2 being clipped, but such an overload is unlikely to occur in normal use. An 18 volt supply gives the greatest overload margin, but in practice a 9 volt supply will usually be perfectly adequate.

Music projects

Microphone pre-amplifier/limiter

Resistors — All 0.6 W 1% metal film

R1,2,8,9,17	3k9	5	(M3K9)
R3,6,14	4k7	3	(M4K7)
R4	47 k	1	(M47K)
R5,22	100 k	2	(M100K)
R7	120 k	1	(M120K)
R10,15,16	10 k	3	(M10K)
R11,12	390 Ω	2	(M390R)
R13	12 k	1	(M12K)
R18	1k	1	(M1K)
R19	5k6	1	(M5K6)
R20	18 k	1	(M18K)
R21	2M2	1	(M2M2)
R23	33 k	1	(M33K)
R24	680 k	1	(M680K)
RV1	pot log 10 k	1	(FW22Y)
RV2	470 k hor preset sub-min	1	(UH08J)

Capacitors

C1,5,10	100 µF 25 V PC electrolytic	3	(FF11M)
C2	1 µF 100 V PC electrolytic	1	(FF01B)
C3,4,7	4µ7F 63 V PC electrolytic	3	(FF03D)
C6	10 µF 50 V PC electrolytic	1	(FF04E)
C8	100 nF polyester	1	(BX76H)
C9	470 nF polyester	1	(BX80B)

Miscellaneous

TR1	BC109C	1	(QB33L)
IC1	NE5534AN	1	(YY68Y)
1C2	LF351N	1	(WQ30H)
IC3	LM13700N	1	(YH64U)
JK1,2	$1/4$ in jack skt brk	2	(HF90X)
	$1/4$ in jack plug	2	(HF85G)

6 Odds and ends

Featuring:

Music projects

Headphone enhancer

When headphones are used with an ordinary stereo (non-binaural) programme source proper stereo imaging is not obtained. The stereo image when using loudspeakers stretches from one loudspeaker to the other producing a sound-stage in front of the listener, but when using headphones the soundstage is from one earphone to the other and therefore within the listener's head! This obviously gives far from realistic results, and the effect can be a little disturbing.

There are ways of obtaining a more spacious effect, and the most simple of these is to reverse the phase of one channel so that the two channels are out-of-phase and fail to produce a stereo image. With this method sounds tend not to have any definite apparent origin, but have a vagueness in this respect. A better effect is produced using a compromise between normal stereo and out-of-phase stereo, with the two channels in-phase at some frequencies and out-of-phase at others. This gives a more vague and spacious effect than normal stereo, but does not totally destroy the stereo imaging.

The circuit shown here is a simple phase shift circuit which can be used as a headphone enhancer. Integrated circuit operational amplifier IC1(a) is a straightforward inverter, while operational amplifier IC1(b) is used as a conventional phase-shifter. At low frequencies capacitor C4 has no significant effect and IC1(b) acts as an inverter so that there is no phase shift through the circuit as a whole. At higher frequencies the coupling to the non-inverting input through C4 reduces the phase

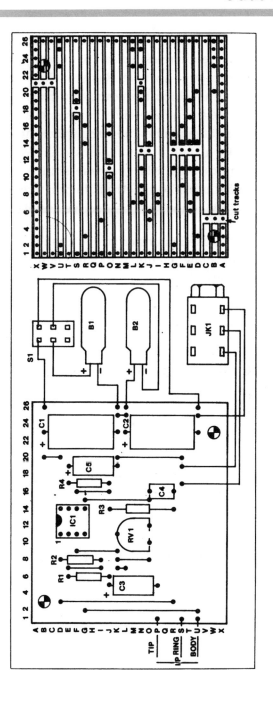

Figure 6.1 The matrixboard layout and wiring for the headphone enhancer

131

shift provided by the circuit. This gives zero phase shift at the highest audio frequencies, and a 180 degree phase shift through the circuit as a whole. Resistor R5 is used to control the point at which the transition from zero phase shift to 180 degree phase shift commences, and with this component at minimum the circuit gives no significant phase shift at audio frequencies (i.e. normal stereo). At maximum value a large phase shift is produced over much of the audio spectrum. In practice this component is set to give the best subjective results.

The enhancer can be used in either stereo channel, and the output will drive low, medium, or high impedance headphones satisfactorily.

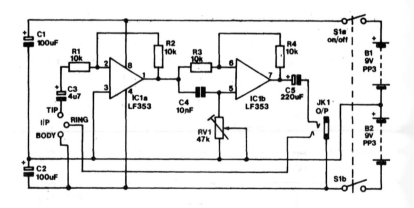

Figure 6.2   The headphone enhancer circuit diagram

The headphone enhancer matrixboard assembly parts list

Resistors — All 0.6 W 1% metal film

R1,2,3,4	10 k	4	(M10K)
RV1	47 k hor sub-min preset	1	(UH05F)

Capacitors

C1,2	100 µF 35 V axial electrolytic	2	(FB49D)
C3	4µ7 100 V axial electrolytic	1	(FB18U)
C4	10 nF poly layer	1	(WW29G)
C5	220 µF 16 V axial electrolytic	1	(FB61R)

Semiconductors

IC1	LF353N	1	(WQ31J)

Miscellaneous

S1	DPDT sub-min toggle	1	(FH04E)
JK1	$^1/_4$ in jack skt stereo	1	(HF92A)
	jack plug stereo plastic	1	(HF88V)
	cable twin (as required)		(XR21X)
B1	9 V PP3-sized battery	2	(FK58N)
	battery clip	2	(HF28F)
	0.1 inch matrix board		
	plastic or metal case		
	connection wire (as required)		(BL09K)

Music projects

Stylus organ

Conventionally a stylus organ uses a simple oscillator in conjunction with low frequency amplitude or frequency modulation (tremolo or vibrato) to produce a more interesting and musical sound than that given by an unmodulated oscillator. This circuit uses an alternative approach and has two audio oscillators. One of these produces the main audio tone signal while the other is a few hertz off-tune and mixed with the main signal at about −10 dB. This gives a much *richer* sound than using a single oscillator and in this respect the circuit is at least equal to conventional designs, if not superior.

The two oscillators each use a CMOS 4046BE device which is a low power phase locked loop, but only the voltage controlled oscillator section of each 4046BE is used in this circuit, plus the internal 5 volt zener diode of integrated circuit IC2 which gives a stabilised supply for the tone generators in conjunction with load resistor R4 and decoupling capacitor C1. A number of 100 k preset resistors are used to provide a series of voltages which give the appropriate notes from the tone generator. Thirteen presets are needed for a single octave organ, and twenty-five are needed for a two octave type (this gives a complete scale including semitones in both cases). Resistor R2 enables the secondary oscillator to be set just off-tune from the main oscillator. The two oscillators will track with sufficient accuracy over a one or two octave range. Resistor R1 takes the control inputs of the oscillators to the negative supply potential and blocks oscillation when the stylus is not connected to one of the tuning presets.

134

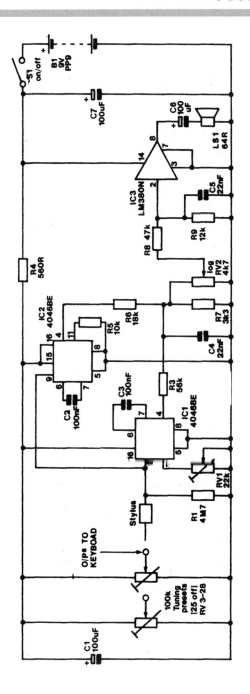

Figure 6.3 The stylus organ circuit diagram

Music projects

A simple passive mixer circuit is used to combine the outputs of the tone generators at the correct relative levels, and the signal is then taken via volume control VR1 to a simple audio output stage using an LM380N. This gives an output power of about 100 mW r.m.s. into a high impedance loudspeaker which should give adequate volume, but an 8 ohm speaker can be used if higher output power and volume are required. Capacitors C4 and C5 are used to reduce the harmonic content on the output signal which is otherwise excessive due to the squarewave output signals from the tone generators.

The unit covers one octave either side of Middle C.

The stylus organ matrixboard assembly parts list

Resistors — All 0.6 W 1% metal film

R1	4M7	1	(M4M7)
R3	56 k	1	(M56K)
R4	560 Ω	1	(M560R)
R5	10 k	1	(M10K)
R6	18 k	1	(M18K)
R7	3k3	1	(M3K3)
R8	47 k	1	(M47K)
R9	12 k	1	(M12K)
RV1	22 k hor preset	1	(UH04E)
RV2	4k7 pot log	1	(FW21X)
RV3–28	100 k hor preset	25	(UH06G)

Capacitors

C1,6,7	100 µF 35 V axial electrolytic	3	(FB49D)
C2,3	100 nF poly layer	2	(WW41U)
C4,5	22 nF poly layer	2	(WW33L)

Semiconductors

IC1,2	HCF4046BEY	1	(QW32K)
IC3	LM380N	1	(QH40T)

Music projects

Miscellaneous

S1	SPST sub-min toggle A	(FH00A)
B1	9 V PP3-sized battery	(FK58N)
	battery clip	(HF28F)
LS1	loudspeaker 66 mm 64 Ω	(WF57M)
	stylus	

MAPLIN Books

This book is part of a new series developed by Butterworth-Heinemann and Maplin Electronics. These practical guides will offer electronics constructors and students a clear introduction to key topics. The books will also provide projects and design ideas; and plenty of practical information and reference data.

0 7506 9053 6 **STARTING ELECTRONICS**

0 7506 2123 0 **COMPUTER INTERFACING**

0 7506 2121 4 **AUDIO IC PROJECTS**

0 7506 2119 2 **MUSIC PROJECTS**

0 7506 2122 2 **LOGIC DESIGN**

These books are available from all good bookshops, Maplin stores, and direct from Maplin Electronics. In case of difficulty, call Reed Book Services on (0933) 410511.

Thugs, Mugs

and Violence

The story so far

By

Jamie O'Keefe

1

Published by New Breed publishing

Reprinted in Jan 2002 (7th reprint)

Printed by

New Breed Publishing

Po box 511

Dagenham

Essex RM9 5DN

www.newbreedbooks.co.uk

A CIP catalogue record for this book is available from the British Library
Printed and bound in Great Britain.

ISBN 0 9517567 8 8

Please note: The theoretical information and physical techniques outlined in this book are for self-protection and Self-defence information purposes only. The author and the publishers cannot accept any responsibility for any proceedings or prosecutions brought or instituted against any person or body as a result of the misuse of any theoretical information or physical techniques described in this book or any loss, injury or damage caused thereby.

In loving memory of my mother who passed away on 20th September 2000 and Reg Kray who passed away 1st October 2000.

*Note:

Nearly 2 years ago I gave my word to Reg Kray that I would never quote him or put in print anything other than what he has given me permission to use. I envisage many people coming out of the woodwork now to profit by the use of Reggie's name, as he can no longer speak for himself. I am a man of my word and will not print anything other that what I printed over two years ago with the consent of Reg.

Dedicated to

My tutor mentor

In life's hidden world

-

My stepfather Peter

(20th July 1929 – 31st August 1993)

Reg Kray telephoned me from prison, after having just undergone eye surgery to talk through the foreword for the re-print of this book.

Due to time restraints and the restrictions that he was bound by, I asked him if he could sum up his thoughts, on this book in a lone paragraph, rather than a lengthy foreword. Although Reg has given me his consent to quote him in length on all the good things that he has said about this book. I have decided to just go with the lone paragraph, which was written by Reg himself. *'Thugs mugs and violence'* had a permanent place within the cell of Reg Kray and was also read for inspiration by the other inmates.

Thank you Reg for you phone-calls, sometimes three a day, to share your thoughts, ideas, opinions and philosophies with me. Rest in peace.

Your friend
Jamie

'Jamie's book 'Thugs, Mugs and Violence' is an insight into the violent times of today and should be read' **Reg Kray – Kray Twins**

Photograph kindly supplied to me for inclusion by Reg Kray

REG KRAY – 32 YEARS SERVED
1968 – 2000 HM Prison. R.I.P.

Acknowledgements

Thanks to: -

My mother and sister for giving their permission to use their details in this book, also for allowing me to use photographs from their own personal collections. Also to my friends.

The late Reg Kray, for friendship, advice, and for always finding the time to phone me and write, also for supplying the photograph of himself and Ron.

Bob Sykes for his early support and help with my profile.

Paul Clifton for recognising my unique approach to Self Protection.

'Martial Arts Illustrated' for helping me to grow.

'Combat Magazine' for a wider audience.

Thanks also to Keith and Anne Richards for original proof reading of this book.

Maurice Heard for final proof reading on reprint.

Roy Shaw for being one great guy.

Kate Kray for her support. (Ron's wife)

Roberta Kray for her support. (Reggie's wife)

My good friend Marty Dee Donovan – West End DJ for designing the front cover of this book and my website.

Melissa for being different & making and old guy happy ☺

Note: Most of the names of individuals and clubs have been changed or altered in some way as to protect the innocent and the guilty. Also to take 'claim to fame' away from the bad guys.

Contents

Introduction

Why on earth would anyone want to read a story about my life?

I certainly don't know but the fact is that I am often getting asked to talk or write about the path I travelled from being a bullied schoolboy to where I am now. That being a martial arts 6th dan black belt, a self-protection instructor, an author of self-defence related books, Magazine columnist, and former Bouncer. Demand has warranted me doing this book so here it is, all in one hit.
I have purposely avoided colouring the book with phrases like *'The wind was blowing with a gentle breeze and the clouds were seeming to smile at me.'* This is a book of fact – not fiction and it is written as one would either speak or write up their own diary.

I have also tried my hardest not to glorify violence by giving a blow-by-blow account of every move performed in the violent encounters contained within this book. However I realise that many martial artists and followers of other fighting arts, will be reading this book for an insight as to which techniques work in reality so I have included a small amount of technical detail to serve this purpose.
I have also intentionally kept the spoken words to a minimum because I don't think its good practice to pad out a book with useless dialogue because you cannot fill it with anything else.
I have tried my best to write it as I felt at the different stages of my life that I am trying to describe, but obvious current reflections on that time period do creep in as I tell my story.
When I am talking about my life as a ten-year-old child, I have tried to talk the talk, as the world would be through the eyes of a child. When I talk of the violence and related areas of my life I also try to tell it as I felt at the time. However the most important thing for me is for you to finally see me as I am now rather than how the press have portrayed me. Also of how I went through the immaturity and ignorance of a child to manhood and how I managed to re-invent myself into the human being that I am now.

All this and to think that I have only travelled 2 fifths of the way through my life - if things go ok for me in the future that is.

Please read this book and understand the affective learning that I have encountered within different learning environments which has made me what I am today.

I did not become a good bouncer because I did martial arts, just as I did not become a writer because I was a bouncer. It was the combination of many aspects of my life that has nurtured the Jamie O'Keefe that you know today.

I apologise, if you are looking within these pages for Jamie O'Keefe, the thug and villain, because such a person does not exist, never has and never will.

This is a true account of my life, which shows that you do not have to accept the hand that you have been dealt in life, you can make changes. It will also, hopefully, make you realise that, no matter how hard a life you may have had, there are many people that have had a far worse life than you. Although this will not make you feel any better, it will show you that you can get through the things that make you feel that the 'end of the world' has arrived, and that you are able to come out the other side, dusted down, and ready to get on with the rest of your life.

Please enjoy my story and learn from the mistakes that I have made but most important of all, gain your own experiences and make your own mistakes. That's an important part of life. If this book inspires you when you're feeling down or helps you think about something in a different way that will make you feel better about your life then my time writing it has been worthwhile.

The true test of the worthiness of this book's existence is if you have trouble putting it down once you begin reading.

Childhood - The beginning

I don't remember too much about the day I was actually born but I'm sure I was a pretty ugly baby and probably the only time I would ever hit the scales at under 16 Stone.

At four weeks old my mother took me shopping with her and left me outside Kentons furniture store in Kentish Town. The world was a safer place then in 1961. After looking around in the shop my mother came out and made her way home leaving me behind. When she finally remembered that she had left me behind she returned to the shop to find me surrounded by concerned passers-by and the police. She said that she forgot that she had a baby. I think that she was drunk or had other things on her mind.

My mother told me that I was a twin and my twin sister was born deformed due to the Thalidomide drug. My biological father demanded that she be put into care where she could be looked after properly and if my mother didn't agree to that he would leave her and take me away.

Me and Karen in East London

I don't know how valid this rumour was because by the time I had grown to any level of understanding it had become a taboo subject that was never spoken about.

One strange thing that I do remember though is that as a child I was taught sign language and was told that one day I would be using it to speak to someone. I don't know if this had any bearing on the story of my twin sister but I never had the opportunity to use it. I can only remember tiny parts of the alphabet now.

So let's begin from the actual furthest that I can recall.

9

I think I was around 9 years old living in Grindal House on the Collingwood Estate in Bethnal Green, East London. A stone's throw from the Blind Beggar Public House where Kray Twin, Ron Kray shot George Cornell some years earlier.

I didn't realise how well known the East End was throughout the world. It was just the place that we lived and as far as I was concerned, the rest of the world was the same. I had no reason to think otherwise.

One thing that I did know was that we were poor. Most people were poor in the East End apart from the villains. The local pubs were full of villain's. It was the way of life.

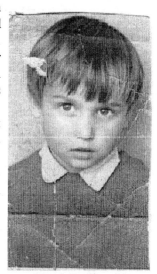

Me in infant's school

My mother used to take my sister and me to Scotland to visit my Nan every time she had a 'domestic' with my father. One time when

Karen - an innocent child

I was about 8 years old, and my sister was 7, we were on our way to Scotland and the train stopped at Crewe for a short while.

My mother got off to get a cup of tea at the station cafe and didn't get back on. The train left. We obviously made it known that our mother wasn't aboard and the train was held at the next stop until my mother caught up with us via another train. It was becoming obvious at this early age that my mother was doing some very strange things, which I later realised was down to her having a drink problem.

I can remember my mother worked as a barmaid. Not the ideal position for a Scottish alcoholic. Any memories that I had of my

Me and Karen in East London – poor but happy

biological father have gone, except one thing that I will go onto in a moment.

For some reason I became the target for childhood bullying. I think

11

Karen at school

it was a bit of a racial thing, being as my biological father was Irish and my mother was Scottish. That is evident in my name.

Jamie is a traditional Scot's name and I think O'Kccfc might be the Irish connection. It was either the racial thing or being poor that made me a target for the neighbourhood and school gangs.

I was never able to speak to my mother about this because she was always at work when I was home and my father appeared to give his time to my younger sister Karen.

Karen was 2 years younger than I was and we fought each other in the way that most kids do when growing up together. I think that she disliked me more though because she once dropped a house brick from a garage roof onto my head, splitting it open and another time when we had a fight she took a dart from the dartboard and rammed it into my neck. So basically I was in my own little world.

Some of the other kids that were bullied would just skip school, play truant and avoid the beatings, but I had to go because I needed that free meal and bottle of milk that the school supplied.

Don't get me wrong, my mother fed and looked after us, but big evening meals did not happen because my

Me and Karen at Grindal house

12

mum would have to leave for work soon after we got home, so it was the usual sandwich or toast.

Sweets were a luxury that we could not afford but my father would always have a bag of his favourite sweets locked in his wardrobe. Looking back now I can see how drawn within myself I became. I was jealous of Karen getting all the attention and realised that she was getting sweets when I wasn't.

All kids go through this feeling, that at some time they are not cared about, but some kids were worse off than others were. They had no visible father for a variety of reasons, from broken marriages to being in prison, so I suppose I was not that bad off in comparison.

Those days we would come home from school and let ourselves indoors by using the key that was on the end of the string inside the letterbox. How things have changed.

Anyway, one day I went home and my mother was out at the shops so I thought that my sister would be with her and my dad at work. I decided to go into my dad's wardrobe to have one of his sweets.

But to be clever, I was going to take the sweet from its wrapper, suck it a few times then wrap it back up. Pretty disgusting eh! But not to a 10 year old craving a sweet. They were toffees so I didn't think it would be noticed. Also I didn't feel so bad because it felt like I was only borrowing the sweet for a little while rather than stealing it. It's strange how you see the world from the eyes of a child. I went in and made my way to my parent's bedroom, but as I approached

Me – the toffee thief

13

the door I heard the radio on and my dad's voice faintly in the background. He was groaning like you would if you had a stomach-ache. I decided it was not a good time to have a sweet tooth. It also crossed my mind that it was not a stomach-ache and that my parents were getting it on or as a 10-year-old would say, *'Doing Rude-ees'*.

I decided to go and see the old lady a few doors away. She would send me on errands for her to the shops and in return give me a sweet. Today she just wanted a local paper and some milk, so off I went.

At the shops I bumped into my mother who had just been to the laundrette. She asked me to help carry the washing home. I obviously realised that it was not my mother and father back home in the bedroom and that my first diagnosis of stomach-ache was the correct one. On arriving home I realised that the only other person in the house with my father was my sister. As naive as I was, certain thoughts did cross my mind that something was not right.

Later that evening I took Karen into our bedroom and asked her if anything was going on, anything that was wrong.

I can still remember the tears filling up her eyes as she said

'Please don't tell, I'm so frightened, daddy will be put into prison and we will be put into a children's home where they will be cruel to us'

'What is happening, I promise I won't tell' I said.

She sobbingly replied,

'Daddy's making me do rude things to him'

Can you imagine that weight of information being declared

Karen going through private hell

to you?

At 10 years old!

I did not realise the seriousness of what was happening, but I knew it was bad and wrong. I don't know whether I needed more proof myself, I had disbelief, I was keeping my promise, or was just plain scared but I don't remember acting on that information straight away. I went into a trance like state for maybe a few days. One day my sister offered me a sweet. She had never done this before but thinking back now it was because she wanted a friend. I also realised that I had been blotting it out of my mind because of the fear of what may happen. The sweet woke me up to the fact that this ugliness was still happening. I told my sister that I'm going to tell our mother. Karen went into a fit, which unbeknown to me at the time was an asthma attack.

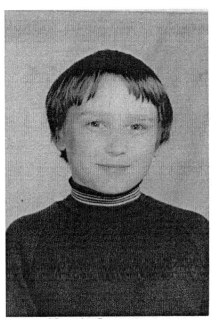

Me – the fire starter

I called my mother to come and help but Karen obviously thought I was going to spill the beans. I thought she was going to die. All this because of my father's wrongdoing. Karen eventually calmed down and fell asleep and we all retired for the night with the secret still eating away inside me.

In the middle of the night I went to the kitchen and got the kitchen size box of matches, went to the living room, opened the box, struck a match and put it back into the box. An immediate flash occurred and I then pushed the flaming box under the foam sofa.

That seemed to be the only solution. Burn the house down, and that I did.

The house burnt throughout, completely gutted. But it solved nothing. We ended up still as a family unit elsewhere. I decided to tell my mum what was happening to Karen. I cannot remember her reaction, all I remember is that the police came and took my dad away and that was the last I ever saw of him.

I look at my own little girl now and wonder how a parent or any adult in fact, can take advantage of a child to satisfy their sexual perversion. I can still see my sister's child like face, sobbing her heart out as she told me her secret. Some 25 years later I wrote my first ever book called **'Dogs don't know kung fu'** which I dedicated to my sister, mother and all others that have suffered from the hands of a male.

While the court case was going on my mother sent us to live in Scotland with my Nan. She lived in Bellshill, Lanarkshire, I think that's near Motherwell.

It was good staying with my Nan because we never saw her much. The down side is that we were a couple of English kids in Scotland going to a local school where the English were not flavour of the month.

Luckily I had many Scottish relatives near to where my Nan lived, so help was always at hand if I was getting picked on too much.

My Nan was also bringing up my 4 cousins because my aunt went off to England and left them, so the house contained 8 kids, my Nan and my uncle John. It was a nightmare but also brilliant. I remember this time of my life with great fondness.

Once a month a parcel would be sent to Karen and me. It contained sweets, chocolate, comics, etc which was great.

We didn't get all that whilst in England.

I don't quite know how long we spent in Scotland but I think it was just for the summer school holidays. I was 11 years old when we returned and remember my mother collecting us with her 'friend' Peter.

It turned out that Peter was my mum's boyfriend but she tried to conceal it until we adjusted to the idea gradually. They met under strange circumstances. Peter was a crook, villain and con-man that

16

would obtain money by theft or deception from anyone he could.

He was drinking in a pub that my mum worked in and overheard her saying that she would like to buy a colour T.V.

At this time colour TV's would have been the equivalent to the Personal Computer now, as far as cost goes.

Peter's game was to steal from the greedy. His philosophy was that if someone is prepared to buy stolen goods, they were greedy, dishonest people and deserved to have their money taken from them. He also believed that if someone parted with their money for stolen goods and got nothing in return, they would not go to the police because they could not explain that they lost their money buying stolen goods.

That's exactly what he did. He would find out what somebody was in the market for, and then offer to sell that item to them at half price because it was stolen.

They would part with the money and never get anything for it. They would obviously feel angry and humiliated, but worst of all they could not go to the police. His confidence trick was called the 'Corner game' in the East End.

I neither approve nor condone this method of obtaining money by deception, I'm just telling it like it was. However if you think about it- have you ever bought something that you knew was stolen? If so, how would you feel if you received nothing in return and could not go to the police? Do you also deserve to lose your money? It makes you think, doesn't it!

My mum turned out to be the next victim of Peter's. He talked her into buying a colour T.V. Because it was going to take quite some time for my mum to get the money together, Peter took his time whilst selling his invisible goods elsewhere. Before the day came for my mum to pay for the telly, Peter had got to know my mum a little better and found out about her two kids that had been sent to Scotland whilst her Husband was on trial for sexual assault.

Somehow she must have touched on the only sympathetic area of his heart and he asked her out on a date. Next thing I know was it had all been arranged for us to come home to England.

The Senior Years

The next stage of my life that I can remember was being accepted for one of the most prestigious schools in London, that being the London Nautical School in South London. I was so exited about it because it was a naval academy, which based all its teachings around sea life. I can remember wanting to be in the Merchant Navy and if I worked hard enough over the next 5 years, I would get in there with no problems.

Peter was making plenty of money from his crooked ways and splashed out on my school uniform. It was a replica naval outfit with the peaked cap with anchor badge with gold weaving on it. It was just like one of those military academies that you see in American films these days.

It was not too long' till I found the 'class' system was in existence within the school. There was the upper class that came from the well-to-do families and below them was the shit. Well you can guess where I came within the scheme of things.

Coming from the East End and having the cheek to dare enter their selective school was one thing, but being poor and having an Irish surname put a few more nails in my coffin.

I can remember being the only kid in the school lining up for free school dinners. There was one line for the rest of the school and another for me. That's no exaggeration. It's exactly how it was. Then when I made my way to the table to eat my dinner I would get pushed and shoved away from any table I tried to sit at. I can guarantee that I was tripped over at least once a week, sending my dinner all over the floor. The teachers were bastards, they had no time for

1970s – Sad clothes, & haircuts

18

supposed East End scum and would make me clean up my dinner, then put me out to still parade in the school playground.

It was the punishment that they loved to dish out prior to the cane. If you had done wrong, in the teachers eyes, you would be put out to still parade from the time of the incident until after school finished. The traffic would pass by and you would suffer all the usual name-calling.

The usual stuff was *'Sailor boy, you look lovely'* or *'Where's your boyfriend.'* In those days a naval uniform was associated with homosexuality, for some reason.

Within school though, there was none of that, it was a 100% macho heterosexual environment, which was fine by me because I certainly was not gay, so that allowed me to be accepted within the school generally.

There was one black kid who came into the school and received the same kind of shit treatment that I did. But things eased off for him after a while when it came to someone's attention that the lad had wealthy parents. They were doctors and would drop him off in their big expensive flash car. I cannot remember the make or model, just that it represented wealth. He also used to bring in bags of sweets and share them out amongst those that befriended him. Eventually he became targeted every time he walked into the school gates and basically got robbed of all his sweets. They were like vultures swooping in on him, stripping him of all his goodies. Many of the school bullies left me alone whilst having a go at the black kid because they could profit from their bullying in a material way, with me, they did it just for fun.

The school sport was rugby, which allowed the bullies to give out an authorised kicking to the victims. I had the studs from the boots stamped into my face many a time and was always getting bundled. The trick was to throw the ball to the victim (me), which I was told to keep hold of, no matter what, or I would get beaten up in the changing room. Once I had hold of the ball I was then like a red towel to a bull. They would steam into me. It happened time and time again until one day the weight of them all on my chest, restricted my breathing and I became unconscious. They shit themselves. Thinking they had seriously injured me.

The games teacher revived me and took me to the changing rooms.
I think he must have realised that if he was doing his role as a teacher out on the field, rather than sitting in the shower room getting off on 11-year-old naked boys, this would not have happened. He promised me that this would never happen again and that I could pick swimming instead to keep away from rugby.

For some reason I associated rugby and all similar sports as a minefield for being bullied. I have never enjoyed or followed sport since that time in my life. I do not know to this day whether the bullying I received on the sports field has had any bearing on that.

Present day. It is quite clear that physical violence is a common occurrence on the sports field as well on the terraces.

I stopped all sports as from that day and decided that swimming might be a good idea, seeing as I intended going into a naval role and could not swim. For my first lesson I was to sit on the viewing area to watch a lesson. This was to make me feel more comfortable with my surroundings. Halfway through the lesson the teacher had to take a telephone call from the school. All the boys were instructed out of the pool until he came back. I was asked by one of the boys to bring his towel to him. I was in the dry area looking after the towels. I took his towel to the poolside and the next thing I knew I was at the bottom of the deep end, fully clothed, drowning. I was gasping and swallowing water whilst also being held down. I can remember a crowd of lads swimming round me like sharks blocking my route. Next thing I remember was being cared for by the teacher. I was coughing and being sick with fear. To this day I don't think anything has frightened me as much as this incident. I thought I was going to die.

It goes without saying that the lads described the incident as me falling in the pool and they all jumped in to save me. Absolute bollocks, of course, it was their way of having fun and another form of bullying.

I had grown to detest the kids in the school and often dreamed about horrible ways to repay them for making my life hell.

I can understand how some kids are pushed to the limits of tolerance and sadly attempt suicide. I loved my mum too much to do that to her and to be honest I don't think it ever came into the equation

All I wanted to do was hurt the bullies, even kill them.

I was 12 years old by then and the thought of killing another human being was going through my mind. I am so ashamed to admit that I felt that way, but, the way I felt, was the way I felt. Beating the bullies up would obviously have been a more reasonable choice but to do that I would have to be able to fight, which I couldn't. I was in a fight or scuffle nearly every day but I was never really fighting back. I would just cover up or try to hold them still until a teacher came and broke up the fight. I was always too frightened to hit back in case it angered them more and they in return, hurt me more.

I so badly wanted to shoot, stab, or kill by any means possible. It would have released every bit of pain and suffering that I had encountered over the last two years at school. I could not face doing another three years or more. In fact, I was growing to detest the people that I was maybe going to end up working with when I left school and started to shy away from wanting to go into the navy. I realised that the upper class and those that thought they were upper class, were not my kind of people.

My mother was not aware of the bullying, or should I say the extent of the bullying, because I did not want to worry her. At this time Peter had been put in prison and she was working hard to keep us, the home, plus send in what she could to Peter. He was able to get extra tobacco and better food by sending money to the friends or family of the screws (Prison guards).

We were struggling so badly. To add to this my mother began drinking more and more. Alcoholism was getting a hold on her. The summer holidays came which was a relief for me. I had grown to hate the school I was at and as each day of the six weeks passed by I was becoming more and more anxious and dreading my return.

As luck would have it, Peter was due for release from prison and my mum applied to the council for a transfer away from the East End to start afresh.

Visiting Peter in prison was a terrible ordeal. So much unhappiness was visible from prisoner, to guard, to visitor. My mum would pour miniature bottles of scotch into Peter's tea for him.

He would always stagger back to his cell half pissed at the end of visit time and away we would go until next time. I was so depressed

21

with my life but hoped things would get better when we moved.

I had to go back to school for a while until the new house came up. Once we knew where the house was to be I could then select a school.

The bullying was easing off a bit because there was a new selection of victims who had just enrolled in the school. However, I still had my usual crowd of bullies that ensured I didn't go too long with out a beating.

One day, after the normal slap around from the gang, I was dragged around to the back of the playground and tied face on to the railings with my tie. I wondered if it was to be the 'Pissing up the back' game or the best greenie phlegm ball to hit the back of my head. Today's game turned out to be a stage further advanced.

Something had happened to do with the IRA, either in the UK or Northern Ireland and a visit was made to the UK from members of the IRA for a peace talk.

This was today's excuse for pissing the bullies off, so it was time to punish that kid in the school with Irish blood in him, and an Irish surname. It was the 'Burn the nylon shirt with your cigarette' game. I couldn't do shit all about it.

They must have burnt about 50 holes in the back of my shirt and they burnt me numerous times in the process. One of the bastards held his fag on me and I screamed with pain. Straight after, I made out to faint and just hung there all limp and lifeless. They quickly supported my body and untied me, then ran off.

I got myself together and just sat there crying, wishing so much that I could learn to fight. After finishing my little tear session I climbed the school fence and went home.

I decided to tell my mother what had happened. I just couldn't go back to that school again. I got home and my mum was crouched over in the armchair asleep with an empty Martini bottle next to her. There would be no talking to her today, she was gone. Karen was making an effort to try and cook herself something. I took over and made us a meal. I left something in the oven for my mum for when she woke up. I went into my bedroom and pulled the covers over my head and sobbed until I was dry of tears and had stopped gasping breaths.

On my return to school I was put on standing parade for the whole day for leaving school without permission. One of the bully gang members came to see me and let it be known that if I told anyone about the ordeal with the cigarettes, I would be dealt with severely. With the message came a forceful crippling knee strike to the gut.

A few weeks passed and it was time for Peter's release from prison, which came at the same time as an offer of a property in Dagenham, Essex. My mother jumped at the chance of getting a house with a garden. I have referred to our current accommodation as a house, but it was actually a flat in a block.

Peter had been released from prison and was not able to work legally, or illegally. The police were on his case just waiting for him to slip up and nobody wanted to employ an ex-convict.

To help out I would take all our unwanted items down Brick Lane and set up on the pavement to sell them. Boot sales did not exist then but I suppose that's what I was doing. I never made much but it all helped whilst no money was coming in.

We had been given a date for moving into our new home so my days at the dreaded torture camp school were coming to an end. One of the teachers announced to the class that I would be leaving the school and that the class should wish me good luck.

That was fine but once the bullies realised that they were going to have to find a new victim, they decided to intensify the beatings as a going away present.

A school trip had come up and the school had paid for my place so I was going to make that my last day. It was a visit to a ship, which fell in line with the nautical thing. It was the usual thing, packed lunch and drinks in plastic bottles only. It was my last day so I decided to pay back the bullies before I went, for giving me so much kindness in the form of beatings. I still could not defend myself physically or should I say did not have the courage and know how, so I did it another way that was cowardly but felt good.

I took some maggots from Peters fishing basket. He used to fish some weekends and had a supply of them. I took them on the trip with me in a plastic bag. The teacher took charge of all the food and bottles of drink, which were kept in a big holdall, so that we would not burst bottles or sit on sandwiches and all the usual things kids

23

do.

Each wrapped sandwich and drink was labelled with each child's name. Well the rest you can guess. I don't remember what reason or excuse I gave for needing to go back to where the sandwiches were but I got there. I put the maggots into the drink containers of the three bullies that were on this trip, and also into their sandwiches, not just laying on them but actually inside the sandwich. I then put a plastic bag in each of the bullies coat pockets with maggots sealed in the bag.

Lunchtime came and the teacher distributed the food. The maggots were neither visible in the sandwiches or the hard plastic drink containers.

It wasn't too long before one of the bullies felt something wriggling about in his mouth and spat it out, he was on his second half of the sandwich by then, and the same went for the other two. They immediately took the lid off their drinks and guzzled it down as if it was a mouthwash that was going to rid them of the maggots they chewed. As one of them received his mouthful of wriggling maggots they were all screaming like mad and being sick. I felt so good. Revenge felt nice. All the other kids were laughing at the bullies. Another kid whispered to me

'I know it was you-nice one, we all hated them three' I smiled.

Apart from that one kid, nobody else had worked out what had happened. There was too much commotion going on. The teacher told us to all check our food and drinks and collected in the bully's spoilt lunch for disposal.

While the teacher was away from the cabin one of the bullies went to his coat to get a handkerchief and pulled out a bag of maggots. As soon as the other two saw it they steamed into him and it all kicked off. Nobody bothered to go and get the teacher. We all wanted to see them suffer at the hands of each other. They hurt each other pretty badly but were finally separated on the teacher's return. The teacher made us all empty out all our coat pockets and guess what?

The other two bullies also had bags of maggots in their pockets. It was so rewarding watching them turn on each other not knowing which of the three had done it.

They were sent before the head and all put on playground standing duty with a teacher present to ensure they did not kick off again. They had been expelled from school and were waiting for their parents to come and collect them. My bus home stopped right outside the school playground where we were all put on parade. It was my last day and I was so looking forward to my last bus ride home. My bus came and I boarded it. They were the older buses, without any door, so I was on the platform as it was pulling away.

I called out to the three bullies,

'Guess what? I did it, it was me, maggots for maggots, see ya, you wankers' I was giving them the wanker sign as the bus pulled away and had tears coming from my eyes with laughter. They were gutted.

That's the last I ever saw or heard from them.

Peter spoke to me when I got home. The school had phoned to ask my mother and Peter if they were aware of any involvement that I may have had in the maggot incident. Peter stuck up for me and denied any possibility of my involvement. He obviously sussed that I did it but would not side with the school. He hated authority and would protect his own family, or in my case, step-family.

He asked me why I did it and I explained everything to him about the bullying and how I kept it secret from my mum. Instead of punishing me, he praised me and said that he will toughen me up.

'There's a way to do everybody,' he said,

'And bullies are no different, but you should not have put maggots in their sandwiches, you should have shit in them'

I laughed, but he was not joking, he was dead serious. He had seen all kinds of bullying in prison and all the possible ways that bullies could be dealt with. He was going to educate me about life. His phrase *'There's a way to do everybody'* originally related to him being able to con anybody. Later in life I adapted and now use that quote in relation to dealing with violence. My quote is *'Anybody can do anybody - you just have to find a way'*

Men use physical violence and weapons and some females do it psychologically by tearing out your heart and playing with your emotions. Anybody can do anybody no matter how big and tough they appear to be. So don't ever think you are beyond being done.

Moving Home

We moved to a house in Dagenham and I was sent to Erkenwald Comprehensive School, which faced Dorothy Gardens where I lived. I think I was about 13/14 years old then. It was so different to the navy school. One major difference was that it was a mixed sex school so that lessened the chances of being bullied by 50% (hopefully). Karen was also in the school so I had someone to talk to until I made some friends.

It was a good school and bullying was not so rife. The thing was to be the best fighter in the class, then the best in your year, and then there was the best fighter in the school. This kind of concept kept the bullies at bay because the best fighter of the class was always ready to take on a bully and beat his brains in.

I decided to become part of a group as a way of mingling in and not to get set upon as I did at my last school. The 60s T.V. program Happy Days had just hit England, causing a following of all rock 'n' rollers. Some even went as far as becoming second-generation teddyboys. Before we knew it Dagenham was flourishing with gangs of teddyboys. Another thing that boosted the movement was the Punk rock explosion. I had my hair all chopped off and made into a D.A. (Ducks Arse) and joined in with the Rock and Roll movement. I thought that it would give me safety in numbers but soon found out that it made me stick out like a sore thumb. Everyone hated the Rock and Roll lads.

I made some good friends in the Rockin' scene but had none outside of it, so decided to hang in there and deal with things as they happened.

I was still being picked on at this new school but things were different. I was picked on because of the haircut I had and the people I hung around with, rather than because of my surname or the political situation in Northern Ireland. I still did not have the balls to stick up for myself and would accept a punch or two rather than chance fighting back and getting hurt even more.

One day a group of lads were giving my sister a hard time in the playground, I suppose I was about 14 then. They surrounded her and were pulling at her clothes, trying to take her blouse off for a laugh. I don't know what made me do it, but I steamed into the lot

26

of them like a lunatic. I did not think for one moment about my own safety, or whether I could fight or not, I just steamed in.

They let my sister go and proceeded to give me a kicking. I lay there all battered, scratched and cut but I felt brilliant. My adrenaline seemed to give me strength rather than the usual weakness that I normally felt. I felt like I was a fighter, even though I had been beaten up. It was so strange how I would never stick up for myself, yet found it in me to stick up for my sister. It was the lesser of two evils brought on by a choice that I had made which made me feel in control. This incident had exposed me to the feeling of bravery. Something that I had not experienced before in this way.

News soon spread around the school about how I challenged the gang and by the time it had done the full circle; I had supposedly challenged anyone who wanted to try their luck against me. I was shitting it!

Every playtime and after school someone wanted to fight me to make a name for themselves. I don't know why because I hadn't even won a fight. All I was good at was taking kicking's. My mistake was to challenge the toughness of a few of the known fighters by sticking up for my sister, by steaming them. I was to be made an example of, or they would be ridiculed.

Around this time problems were beginning to develop at home. My mother and Peter had married in July 1974 and I let my mother know that I disapproved of this. Peter was rapidly changing from the nice guy image that he portrayed when we all came together as a family unit. He was slowly becoming a monster. It was the

The wedding – Barking 1974

old classic tale of the outsider coming in to the family and grooming by buying their way in with gifts or niceties, and once they were a certain stage into the set-up, their real side would emerge and the grooming would be replaced with brutality.

He would drink continuously and become more verbally abusive towards my mother. It did not take long before the insults turned into strikes. It was the Jeckle and Hyde syndrome.

Unbeknown to my mother, he was also directing his heavy-handiness in my direction.

He hated my biological father for obvious reasons and saw aspects of him in me once the drink took control of him. To my knowledge he had never seen him, but after he downed a bottle of scotch I was the recipient of his abuse.

I knew that had to learn to fight in order to protect myself and my mother from his abusive behaviour. I also wanted to beat up a few of the kids in the school who were giving me a hard time.

The school took some of the kids to Judo classes but I was denied this choice because I was getting into too many fights within school. So I joined a private club, under a guy called Dennis Darlow. I think it was in Stratford, East London or somewhere near to it. I liked Judo and took my 5th Mon, which is a junior grading, in Nov 1974 at Redbridge with the British Judo Association but I desperately wanted to learn something that had kicks and chops in it. That would make a good fighter. I asked my Judo instructor if he could teach me any of the Judo chops. He explained that Judo was a sport and that there was no such thing as the Judo chop. Movements like that came from Karate. He secretly taught the class a series of Karate movements, which he said was called a Kata. I loved it; Karate was what I wanted.

I wanted to be able to use a chop or kick to lay out anyone that wanted to fight me. Also I was determined to learn enough to eventually pay back Peter, who was now my stepfather, for the harm he was inflicting on my mother and me.

Around this time the children's T.V. programme **'Blue Peter'** featured Karate on it and had this boy on who was one of the youngest Karate black belts in Britain. I recognised this kid as a pupil in my school. His name was Lenny Johnson and he trained with Ticky Donovan in Dagenham. I think it was around 1975, when I was 14, Lenny would have been 13. I was not aware of how massive Ticky was within the Karate world. Ticky was a 4th Dan at this time and he had won the world karate championship 3 years running.

I made friends with Lenny and asked him millions of questions about Karate. The martial arts boom was just happening and I wanted to learn it all so badly. I got myself a paper round so that I could save up for Karate lessons. I was still doing Judo, but did not value it at all because it was free and readily available. I was so naive.

To think of the impact the grappling arts have within the martial art world now.

There was a shop in Barking called 'Ron's dojo' which sold martial arts equipment and also had a small training room upstairs. I used to hire it for myself and train in what I thought were Karate techniques.

One day I met a guy there called Neville Sheen who was a black belt in Ishinryu Karate, Ticky Donovan's system. I asked him if he would train me privately in Ron's Dojo. We worked out a price, that was affordable to me, and training began. I told him that I was not interested in the sports application of Ishinryu, I just wanted to learn how to fight. In fact that's why I didn't join any of Ticky's classes, it was all sport. Brilliant stuff, but that wasn't going to help me in a real fight, and I mean this in the most respectful way.

Neville was a hard looking man with a scarred face, the type of person that I wanted to be like, respectful, polite, but hard. He would always turn up for training and gave me a good workout. He would even stick in some boxing to prepare me more for real fighting. This was the best period of my Karate days. I was a Karate white belt, just full of enthusiasm, I lived and breathed Karate.

After about 6 months of private lessons from Neville I went on an Ishinryu Summer camp at Clacton with Lenny Johnson and a brown Belt called Simon Kidd, it was June 1976. Simon is a big name in Ishinryu now. I trained solidly for the week with Ticky and all the other instructors and received more training from Lenny and Simon in the evenings. At the end of the course I took my 6th Kyu Red belt in Ishinryu. I was so proud, it meant so much to me. I now felt officially part of Ishinryu and also a Karate student who was part of something bigger. Simon's dad, Fred, collected us at the end of the week and dropped me home to my mum's in Dagenham. My mum was so proud of me. I actually felt like I could fight now. The certificate is still in my possession now, some 25 years later. I continued my lessons with Neville and progressed through a few more grades, but for some reason that I cannot remember, my training with Neville came to an end. I think he had a new job or had moved away from the Ilford area and was no longer able to continue teaching me.

I joined Ticky's club in Wood Lane, Dagenham.

It was in an old disused air raid shelter; at least I think that was its original purpose. The atmosphere in the club was electric. If you have ever seen the film 'Enter the Dragon' when Bruce Lee first arrives on the island where the tournament was to be held, he was greeted by the continuous rhythm of the Karate guys doing the punch and Kia, approaching Ticky's club gave you that same sensation.

There was sweat, noise, aggression, and all the things that made the atmosphere great. Unfortunately though, there was also a lot of bullying. Students who were climbing up the coloured belt ladder were treating those of a lower grade like shit. To this day, I don't know if the instructors were aware of this, or even if they cared about it. After taking as much bollocks as I could from some idiot, who did everything to ridicule and make a fool of me, just because I was a much lower grade, I decided to leave.

I was being picked on at home and in school and certainly wasn't prepared to work early morning paper rounds just to pay someone to bully me.

Back at school I told Lenny that I was leaving Ishinryu because I didn't like the way that attitudes were developing within its club. It was either creating or attracting people with a bad bullying attitude and it made me not enjoy my training. I didn't want to be part of that.

I decided to go back to private lessons and approached an instructor who I had met a few shops up from Ticky's place. This was Tyrone White, who was another mega name within Ishinryu.

Ticky let us use his mini Dojo that was situated at the rear of his Car sales showroom in East Ham. Tyrone was a brilliant instructor and

Ticky would also come in and fine-tune my techniques and even teach me if Tyrone could not make it. This carried on for quite some time.

Fights in school were becoming more dangerous because the boys were getting bigger and stronger. I now found myself in the position where I had kicks and punches that I believed would work, but still didn't seem to have the guts to use them. The fights always ended up in a grappling situation with me underneath, reverting to Judo to try and hold the attacker still until he gave up hitting me, or a teacher came to save the day. I was not using Judo wisely, because at the time I quite wrongly did not place any value on its effectiveness.

Plus I wanted Karate to work for me but was still unable to launch a strike. Looking back, I now realise that I was waiting for a certain type of punch to come my way, so that I could block it and then counter attack, but real fights didn't work like that. The school playground bully didn't know how to throw this technically correct punch because he had never trained in Karate. His punch just came from an angle that I was not trained for, at a speed and with power that was unorthodox, plus he made contact with the blow. Also to make things even more awkward we would end up rolling about on the floor with him just trying to hit me as I tried to protect myself. This is not how I wanted fighting to be. I wanted to be able to use my techniques but the awkwardness of reality messed me up each time.

Karate was giving me brilliant techniques, but something was still not working for me. In front of the mirror I could kick & punch, snarl, look aggressive and kia (scream/shout) in a frightening manner, but I still had no bottle. I felt brilliant whilst training or acting out fighting in the dojo, but that feeling disappeared after a night's sleep, and the next day I was still a wimp. What was wrong with me? I knew I could do the techniques effectively, but just lacked the courage to let loose with them.

Although my stepfather was bullying me, he didn't see it as such. He called it toughening up. He wanted me to feel fear so badly that I would kill in order to end the torture. He wanted me to feel what 'Real' bulling was like in the real world, if you were institutionalised, put in prison, where there is no one there to protect you. You just deal with it or die! I don't quite believe his reasons for bullying, I think he was just showing his power and control over our family, however I did learn a hell of a lot from him over the years. Over time it dawned on me that I was setting myself up to deal with a different type of bully in life. Not the school playground bully, who will basically give you a fair fight, but rather I was preparing for the monster that hides inside many men and uncontrollably released all its ugliness on a victim that dares to challenge its power and control. I was gearing myself up to sort out my own stepfather. I could not avoid him or hide like I could the school bully; he was there every day and night. I hated him so much. The frightening thing about it all was that my thought on how to deal with him surpassed Karate movements; I was actually plotting ways to kill him. You may find this a bit extreme but I can remember the feeling so strongly. Many, many kids have thoughts like these, temporarily, go through their heads. I was no different.

One night I lay there crying my heart out whilst listening to him bashing my mother's head into the solid fireplace. She was screaming in pain for him to stop. He shut her up by smashing a heavy tabletop glass ashtray into her skull. That's the kind of bullying you could not run away from. My mother never called the police, that's the way it was in those days. You didn't invite the police into your home for anything, no matter what.

I went to school the next day so full of sadness. I felt sick with fear and hate.

An incident happened in school, I cannot even remember what it was about or what was said, just that it was one of the school bullies giving me a hard time. I exploded into a psychotic lunatic, screaming and shouting at the bully to come and have a fight with me.

I cannot remember one single word that I screamed at him or how aggressive my body language was, but it was psychotic. I totally lost it. The bully shit himself and backed off cowardly in front of all his mates. I taunted and challenged him again but he was not having any of it. I had won my first fight without fighting. I didn't lay a finger on him. From that day on he tried to say hello to me as we

32

passed in school but I was not interested.

Unfortunately for him that day, his 2% of bullying received 102% reaction because of the abuse I had seen at home. The anger and aggression that I wanted to unleash on my stepfather found its way to the next bully in line.

A few days later when things had calmed down at home I spoke to my stepfather about the fight I had won, but didn't dare mention that I was angry because of his abuse towards my mother.

He explained to me that most fights could be and are dealt with in this way in prison. You would psyche someone out by instilling fear as much as possible and this would make most people back off rather than test fate. He said that that's why people are fearful of mentally unstable people.

Not because of anything that they have done, but rather because of the way they act and talk. It freaks people out.

He was right; body language with aggressive or strange verbal dialogue did freak people out. He also told me that most bullies would shit it and back down if challenged. Its not something that they are used to or know how to deal with. I took on board all that he had to say but it still did not change the way I felt about him. What you feel - is what you feel, be it right or wrong.

News had travelled round the school about how I had beaten up the bully, which I hadn't; I had confronted him - that's all. Now I had a mixture of people wanting to be my friend or wanting to fight me to show what a wanker I really was, and I was.

Fighting now became an everyday thing but something was changing. Due to the sheer number of fights I was having, the numbers game dictated that a win had to come sooner or later. I was actually winning some fights but it seemed that the only fights I was winning were the ones where I actually threw the first punch, kick or head butt. I never liked doing this because it went totally against the principle of Karate or the martial arts, as I understood it at the time. I was to wait for a strike to be launched and then defend. There was nothing anywhere about striking first. It was kind of considered as dirty fighting.

The only reason that this came about was because now and again the psychological mental act did not work and the bully would just stand there and say,

'Go on then- throw a punch'

So I was in the position of doing it or being ridiculed. Karate worked wonderfully. I would launch a kick or a punch and that would be it, the fight was over. It was so piss poor easy, that I didn't look too far beyond the single strike.

At that time period around 1975-76 people still didn't know too much about the martial arts and it did scare most of them. They didn't want to mess with it. What also helped was the bullshit that surrounded the martial arts, like the death touch and one punch-killing blow. I rode on that bullshit wave for a long time to help keep bullies away. I found that my sister was also getting as lot of mileage out of it, by using my name whenever she had a problem.

I was actually beginning to make a name for myself, which felt good.

I only had two close mates in school, Peter Chamberlain and Robert Peters which only came about because they were both into the Rock and Roll movement that was going on plus both had different home lives to most other kids in the area. This kind of acted as a magnet and pulled us all together.

Bob (Robert) was a giant of a kid, totally passive without an aggressive bone in his body. He was a great laugh and handy to have around. His pure size would keep people away from me. Peter on the other hand was a roughneck. He was trouble and a bully where he could get away with it. I can remember once he took an old 78 record to school making out that it was valuable. He threw it to another kid in the school playground, knowing that he could not catch it. It hit the floor and shattered. The old 78's were very fragile. The kid who failed to catch the supposed valuable record then had to pay Peter money every week, until the record was paid for.

Another thing that he did was to get another kid who wanted to be part of our group, bunk school with him. They would then go round to this lad's house to hide. Unbeknown to the boy, Peter would also help himself to the food and money that was lying around. Those days everybody used tins and jars to save money in.

For some reason Peter and me began to clash, I think he was starting to take liberties with me and I didn't like it. We parted company.

I spent most of my time with Bob and his brother, but that also attracted its fare share of trouble. One day we were hanging around, doing nothing in particular so decided to pay a visit to Bob's uncle in Rainham. As we were waiting for a bus to arrive a gang of skinheads pulled up in a car, jumped out and gave us a right kicking. I was smashed in the head with a homemade ball and chain style weapon with nails sticking out of the ball. That opened up my head nicely. Then they busted my wrist and arm and finished off with an array of kicks and punches. This ended in a trip to the hospital to be patched up. The police took statements but this is as far as it ever went. I could have avoided ending up with my arm in a sling and my head stitched up if I had run, but that was something I stupidly refused to do.

What I thought I was going to do against five blokes, I don't know.

So if you are going to take note of any advice I have to give, running is acceptable. Do as I say and not as I do. Run away from danger as and when you can. A pair of heels can be worth two pairs of hands at times like this. Let them hurt your pride rather than your body.

Times were becoming harder at home. My stepfather Peter was earning again and bringing all kinds of trouble to the door and his drinking was getting heavier. My mother was now hitting the bottle continuously to blot out the unhappy home life we all had.

I could take no more and left home.

I went to live at Bob's house, with his parents, for a while until I eventually managed to find a squat, in Limehouse in East London.

It was a condemned block of flats that attracted all the 'down and outs.' I was a child in an adult's world, it was horrible and frightening. Somehow, my parents found out where I was and came and took me back home. Everything was fine for a short while but then things became even worse than they were before. Peter seemed to hate the fact that I tried to be a man and look after myself. He set out to show me that I was nothing. He was the only man in the house and I would be reminded of that every time I looked to challenge his position.

I found myself a job on a greengrocer's stall and made enough money to get myself a bed sit in Forest Gate, East London. Then that was it, I was off again. I left the greengrocers so I could not be found, got myself a job in a bakers and was now a man of the world. Obviously due to my home situation I did not take any of my school final exams, but fuck it, I had money and a home and nobody to push me around.

I was now rid of bullies in my school and home life.

I used to hang around at a youth club in East Ham, which is where I met up with two of the funniest guys I have ever known in my life, Ricky Meakin and Steve Borg. Steve was Karen's boyfriend and due to his crazy actions he later adopted the name of Crank. I used to get down there almost every night to meet my mates, and my girlfriend would also make the journey to meet me there. She was my first girlfriend. She was 15 years old and would come to meet me at the club; we would just chill out having fun, listening to music and all the stuff kids did then. Kids are slightly different these days. One day, I was waiting outside the Hartley Youth Centre to meet her as she arrived on the bus. They were the old-fashioned, double decker, platform buses with no doors, you would just hop off or on the bus and pay the bus conductor as he came round to sell you a

Karen in 1979

ticket. These days you have a bus driver and ticket seller all in one with automatic doors so that people cannot jump on an off the buses. This puts a stop to the fare dodgers. Anyway, the bus was just coming to a stop and my girlfriend must have mis-stepped or lost balance as the bus was still moving and she then fell into the path of the car behind.

She was killed instantly. I wasn't allowed to go to the funeral. It was just a family thing, which I

Me, Ricky M, & Meathead 1978

36

understood. That was my first ever experience of having somebody that I knew, die. I cannot remember what impact or effect it had on me but I do know that dating girls was not a priority for me for some time after that.

Training was the main thing that I was into. Not only was I getting very good at it, but I actually enjoyed it when it kicked off so I could be like the superhero that saves the day. I was such a wanker but that's how it was. I still never instigated fights but always seemed to be involved in them. Karen was still at home and became involved with the local Becontree skinheads and some how I became friends with the two main ringleaders, Dave L and Martin C (Meathead). The problem with this was that my associates within the Rocking scene did not like this at all, so began laying pressure on me to break links with the skinheads. At that time and to date I have never liked the skinhead movement but I also did not like being told that I had to break links with my mates and with my sister because of the clothes they wore. She was a skinhead girl with the feathercut, two-tone suit and loafer shoes.

One of the guys giving me a hard time was my old school friend Peter C who I hadn't talked to for some time. Peter stole a door key from Karen and told her to send me round to get it. I would have to fight him for it. I didn't want to fight Peter. I was always slightly intimidated by him and this made things even worse but I knew that if I didn't get the key back for Karen, she would get a good hiding from my dad when she got home. So I chose the option of me possibly getting hurt, rather than Karen guaranteed to get punished. I remember riding over to Dagenham to where Peter C was. I must have been 17 years old by then because I had a Suzuki 50cc moped. I knocked on his girlfriend's door and asked for him. She gave me loads of abuse and then called him out. He looked at me as if I was nothing and said,

'Take your crash helmet off so I can punch fuck out of you'
'I've just come for the key, I don't want to fight' I said.
He threw the key on the floor and said,
'Next time I see you in the street, I'm going to kill you'
I picked up the key and went. He didn't really want to fight me and I didn't want to fight him. His girlfriend caused the whole situation.

I knew that this situation could arise again so trained even harder in preparation for it. I was gradually moving more away from everyone involved in the Rocking scene and was drawn more towards the Two Tone Ska music scene and movement. As the skinhead movement began to follow Sham 69 and become more involved in race wars, due to various right wing groups involving themselves with the movement, we went in a different direction. The punk rock explosion had arrived and with it a new band called The Jam. They were basically playing raw punk style aggressive guitar music but creating the mod revival.

Paul Weller, Pete Chamberlain & me 1979

The mod thing took off in a big way, mainly helped by the 79 film 'Quadrophenia'. Little did I know it but 1979, probably had the biggest impact on my whole life.
I had finally found a sense of belonging, I wanted to try fitting in with something that had been around for

Me in the mod revival of 1977/1979

38

years and already had its leaders, gangs, etc. I was there to help and create a revival scene that I wanted to be part of. I'd like to believe that we were largely responsible for the fame and progression of The Jam and Paul Weller, but as Paul quite rightly recently reminded me, a gang of mods from Dagenham were hardly capable of being the reason behind the Jams global success and that of Paul's which has spanned over two decades. It was down to their own hard work and talent. I had to agree with him but we were still there when the Jam didn't have such an audience so I still feel we contributed in some small way. We were the revival. Although I say revival, it wasn't really a revival for us. It was the first time round for us.

The early Jam days consisted of skinheads and punks who would fight non-stop. Luckily though, Jimmy Pursey and Sham 69 attracted most of the bovver boy skinheads away to their gigs leaving the 'mod' skinheads or suede heads alone to battle it out with the punks. The atmosphere was electric and I actually began to get exited at the thought of violent situations occurring. We went to gigs just to kick off, it's as simple as that. It sickens me now to look back at what I was and what I represented, but I'm sure everybody has skeletons in their closet and this is mine. We loved the music and it somehow used to power us up to kick off. It got to the stage where we would fight anybody and everybody that crossed our paths.

It was never because they were punks, teds, skins, blacks, whites, soul boys etc. We would just provoke anyone and everyone until someone took-up the challenge. One thing that we didn't do though was bully.

We didn't pick fights with those that didn't want it. They were no challenge for us. We wanted to kick off with groups that thought they were hard, so that we could prove we were harder.

If we knew it was going to go off because someone wanted it, or we wanted it, we would set it up so that we would launch our attack as soon as a certain part of a song played. We would always change the song we kicked off to because if it were know by either the band, or the DJ, depending of the type of venue we were at, they would not do that song, plus our rivals would get to know our method of attack. If, for instance, we were at a Jam gig and say our chosen song was 'Private Hell', we would hear the song begin and immediately position ourselves on our mark. That being who ever it was that we each individually wanted to kick off with. It was just like a football team covering the opposing team's players. The difference with us though was that our opposing team did not know that we were plotting up and about to kick off. They couldn't even see us as a mob. We were scattered all over the place and just looked like individuals. Our trigger would be the word *'Smash'* and that would signal us to launch an attack. So as the song was playing it would set our adrenaline into a rush as we waited for our trigger word and then it would get to *'Alone at six o clock you drop the cup, you see it SMASH.'* Then bang! We would go. It never failed because the opposition were totally engrossed in the band, singing along to it with their brains totally engaged with the entertainment that was going on.

If the fight did not finish inside a gig, we would carry it on outside on the tube platforms. There is no doubt about it, as 'The Glory boys, we were right bastards!

One of our greatest tricks was confusing someone before we spanked them. It nearly always finished the fight there and then, in a few seconds. We would suss out who the leader of the opposing gang was, or the one that looked to be in charge. Then one of us would approach him and say something absolutely stupid like,

'Do you think it's gonna rain chickens if the coach fly's tonight?' or, *'Do you think that your garden helps whales?'*

They were just the first stupid, illogical phrases we could think of with the sole purpose of confusing the person we were about to hit.

It was a little trick that my stepfather Peter picked up in prison and had passed on to me, which I shared with my mates, because it worked brilliantly. He taught me that the brain could only think of one thing at a time. If you give someone something to think about, especially if it is something stupid that takes a few seconds to digest; they will not be able to think of other things, like hitting you or you hitting them. It always worked. As soon as I threw a stupid brain 'engager' at them, they almost froze in thought for a few seconds, and then I hit them. Just one bang on the chin and they would go down like a sack of spuds. I had over 300 fights before I even left school so I've had a couple of people do a similar thing to me and lay me out just as easily so I also know how vulnerable we all are to this. Most young people would fill their diary of all the sexual conquests that they had as part of their growing up, however I was less than successful within that department whilst at school so all I had to write about was my fights. If you want to read more about these little tricks that I would use when dealing with confrontations, read my book 'Old school-New school', it goes into more detail. No doubt you will have already seen another version of the brain engaging technique, which goes something like this,

'Oy! What you looking at?'

Or one that my son Adam threw at me lately was,

'Dad, how do you confuse a stupid person?'

'Er I don't know Adam, how do you?'

'Elephants!' he said.

Elephants could have been any other word at all. It was the brain engager. He is learning well!

Teenage Meets Adulthood

We soon become unofficial self-appointed guardians, security, and bouncers for the bands we were following, mainly, The Jam, Secret Affair, The Purple Hearts and the Chords. The mod scene was becoming firmly established and we were becoming a firm to be reckoned with. Fights had surpassed the school playground push and roll around confrontation. Things were much more serious. Weapons were now order of the day and fights did not end when the other person submitted. They ended when bones were broken or blood was drawn. It sickens me so much now to look back at the level of violence I enjoyed. At 17 years old I had gone from being a victim of bullying, to someone who enjoyed the exchange of physical violence. We were a pack of wolves from the East End, known as 'The Glory Boys', who became feared throughout the U.K. mod scene because of our reputation for fighting. Individually most of us were tossers who couldn't hold our hands up in a one to one, with a competent fighter, but as a pack of around 50 the added security made us do things that we would normally shy away from. I would say that many of the group were also linked with the infamous I.C.F. (Inner City Firm) due to a combination of being West Ham supporters and coming from Dagenham, Canning Town, East Ham, Manor Park, Stratford and other surrounding East London areas, plus being former skinheads.

I was, unusually, never into football so did not subscribe to or become involved in the I.C.F. in any way. In fact I have never been to a football match in my life.

In our own little world, that being the mod scene, we were the bollocks. Nobody fucked with us and got away with it. Outside of that world, without the security of the rest of the pack, we were nothing. A stupid bunch of immature wankers, but we couldn't see that, we were ignorant teenagers who just didn't give a fuck. One of the things that made us infamous was the fact that we were getting mentioned in almost every issue of the music papers N.M.E. (New Musical Express) and Sounds, compliments of a reporter who worked for them, who was very into our scene and followed its progress avidly. You will probably recognise him, his name is Gary Bushell, who went on to become a columnist for the Sun daily paper and now appears in many different TV shows. I even remember him putting a picture of himself and Hoxton Tom in 'Sounds' with the pair of them posing as the Krays. The last time I spoke to him was at a party I went to with Glen Murphy (George in London's Burning) and Terry Marsh (former World boxing champion), which Glen had organised to promote his film 'Tank Malling.' We chatted with great fondness about the old days at the Bridgehouse pub and live music venue in Canning Town and how the music scene at the

Me, Terry, and Glen – Ruskin Arms boxing club

time had a big impact on many people, including us.

I was going to be making a self defence fitness video with Terry Marsh and Glen Murphy around that time. I was going to do all the martial art and self-protection stuff, Terry was going to be the fitness man and Glen was going to be the pretty boy. Terry and myself were as ugly as fuck, so Glen was the obvious choice. It was going to be sold straight through on the shelves of WH Smith and I would have made a nice £30,000 for myself. Unfortunately after I had filmed my stuff, Terry Marsh got arrested for allegedly shooting Frank Warren and spent the next year in prison, awaiting trial. Apart from the fact that nobody would touch our project with a barge pole after that, it also made things uncomfortable in the circles we moved in. The police paid me a visit and went all over my house in connection with the shooting, because Terry was in training with me and I was also in possession of firearms and ammunition. My gun and ammo had no relation to the stuff that was supposed to have been used to shoot Frank Warren so they left my home happy that I was not involved in any way. The other bad thing about it was that one of my other mates, Mo Hussein, was on the TV by Frank Warren's side just after he had been shot, in Barking by the Gasgoine Estate, so prominently featured throughout this book. Mo was the former British Commonwealth champion for two years and was involved with Frank Warren. Whenever around each other's houses training or chatting we never spoke a word to each other about the incident. I was just in the middle of it all. After that myself, Glen and Terry all went our separate ways. I have actually known Glen from my days at the Bridgehouse in Canning Town. Glen's dad ran the pub and Glen was just beginning to make a name for himself, as an actor, in a programme called 'Shine on Harvey Moon.' He also had a short musical career when he released a single called 'Jab & Move' for the Repton boxing club.

So as you can see a lot happened in and around the Bridgehouse. The main places for our gang to meet were either at the Barge Aground in Barking or The Bridgehouse where we would arrange which gig and band we were going to see.

One day we heard that there was a gang of northern mods coming down to give us a seeing to because they did not like our reputation. We had no problem with Northerners, or anyone else for that matter, but a challenge was a challenge.

We put the word out that we were going to see the Jam at Aylesbury and that we didn't want a fight. We just wanted to see the band. We hired a coach to take us there, all 45 of us, and planned all the way there how we would do the Northerners. When we got there the Northerners were already there. You could spot them a mile off with their Union Jack targets on the back of their Parker jackets and fur around their hoods. Some even had northern mods painted on their backs. When they saw the extent of how firm handed we were, they shit it. The last thing they wanted to do was get a good kicking miles away from home. They wanted to be friends, so we let them

44

We allowed them to buy our beers all night and just took the piss out of them. Near the end of the gig we were having such a laugh with them that fighting was the last thing on our minds. Next thing we know there are about 30 skinheads outside trying to get in to the gig to hand out British movement leaflets. As soon as they saw that there were some black mods inside it kicked off. They steamed into a couple of these northern boys and their mates ran off. That was our signal to kick off. We steamed the skinheads and kicked right off with them.

Most people are easily intimidated at the sight of skinheads because

Skinheads from Becontree, Barking, Manor Park, & East London

they do look hard. However being as most of our mob were former Becontree and Manor Park skinheads, we didn't give a fuck.

We absolutely beat the shit out of the skinheads. The police arrived and nicked the skinheads, because they were trespassing. We all got away with it, although a few of us did get hurt.

One of our gang had told the Northern mods that another backup mob of skinheads were on their way to get revenge but we wouldn't be there to help out because our coach was leaving soon. But if we paid our driver £100 he would hang about for as long as it took for the northern boys' coach to arrive, to take them to safety. The £5 notes couldn't come out of their pockets quick enough. They gave us the £100 for the driver, which we obviously cut up ourselves, and then gave us all another £10 each for our inconvenience. We had a right result. Lets face it, if we had not helped them out against the skinheads, they would have received a right beating.

The night ended without any more trouble and the northern boys got their coach home and we headed back to East London.

A few weeks after that my stepfather was arrested and put in prison. He hit the headlines nationally in 1979 when an old working partner of his, Billy Amies, turned super-grass. He had basically been offered a lenient sentence for a crime he had committed, if he admitted to all the other crimes he and others had been up to over the years.

Peter suited up for the 'corner game'

Some of these things were robberies, firearms, safe blowing, in fact any crime you could think of. Obviously my stepfather did not expect to be super-grassed by one of his former partners in crime.

I think every person in Dagenham must have known why Peter was arrested due to it being plastered throughout our local Dagenham post and East London Advertiser and the nationals.

It was for a jewellery robbery they did in 1974 in Brighton.

It was around the same time my mother and Peter got married in Barking registry office.

At home, Dagenham, 1974

I was a chubby 13-year-old at the time.

While Peter was away in Wandsworth Prison awaiting sentencing, I

46

went back home to help look after my mum. It was 1979.

My sister Karen had developed a drug problem, which just seemed to happen overnight. At first she was taking French blues, which were tiny blue speckled tablets that were in fashion at the time. Basically they were speed, same as what sulphate is taken for now. This, in itself, did not seem much of a problem because many people in the mod scene were doing it to copy Jimmy in 'Quadrophenia.' However, my sister was still involved in the skinhead scene and the skinheads began a new craze called glue sniffing. So, I'm back home now, my stepfather is in prison, my sister is out of her skull due to her glue addiction and my mother is drunk all the time. I did not even smoke cigarettes, let alone do drugs. I think the martial arts kept me on the straight and narrow path. All I would do is train and train. I had moved away from Ishinryu by now and was training in every other local Karate club that I could find. They were all essentially the same, just using different style names. Eventually I settled with Shotokan, which I felt took me a step nearer to traditional Karate, as it originally was.

Unfortunately my instructor was a bit of a tosser and did many things that I did not think that a Karate instructor should be doing. For one, he was shagging his female students and in turn they were passing gradings. Some of them were barely over the age of consent and he was being trusted, by the children's parents, to educate their children. He was abusing his position of trust.

He would make up stories about being banned from various martial arts associations because he had killed someone with his bare hands and always made out he was carrying firearms.

All the students hated it when a new female came to watch the class because we all knew that he would show off his Karate skill at the expense of one of the students. His karate ability was good but his attitude was bad.

One day it became my turn. Two local girls had heard about this Karate guy who could teach them to fight and also would shag them. Young girls who wanted to be shagged by a man, rather than a 17-year-old boy, were commonplace. I don't know why, it's just the way it was. You just had to be a black belt. I was a 1st Kyu brown belt at the time, well overdue for my black belt but was being

47

held back by my instructor because he couldn't make the grade for 2nd Dan. So he, in turn, held me back.

The demonstration began and the instructor went through the usual display of Karate moves known as Kata, then on to the obligatory board and tile breaking. Next it was the sparring, which he picked me out for. Sparring was just a game, in which no real contact was made, with the purpose of showing the public the skill, speed and techniques of Karate. My instructor began showing his stuff and I was just staying on the defence. This was something that you would do, as a sign of respect towards your instructor. Even if you knew you could beat the shit out of them, it was not the done thing. A minute or two into the sparring the instructor was getting a little carried away with himself and began to throw a few heavy shots in at me. This did not bother me, because I had also started training in full contact Karate elsewhere, so didn't mind the contact. As the blows were coming in harder and faster, the female onlookers were lapping it up. He saw this and stepped up the pace.

He did an axe kick, which essentially meant swinging his leg up as high as possible, then driving his heel in a downward direction towards my crown or face. As his heel came crashing down into my skull, sending me down to the floor, I realised that this was no longer a game. I got up, bowed to my instructor and walked off to the changing room, leaving him to soak up the glory.

After the demonstration was over he came to the changing room. I was still there just sitting and thinking about what had just happened.

He said to me,

'Jamie, that's the difference between a black belt and a brown belt, you are my student and will always be and should think yourself lucky I didn't give you a right beating out there' I said,

'But why? I'm just here to study Karate, this is not what I see Karate as being all about' He replied,

'I heard one of the class say that they thought that you could do me and I cannot have my students losing respect for me'

I couldn't believe what he was saying! I said,

'This is not the club for me, and you are not what I want from an instructor'

48

I then got up and walked out.

Later that night I had a phone call from one of the other students saying, Sensei wants to meet you at 8pm tomorrow night at the waste ground, or alternatively, you can write a letter to him stating that you were to scared too accept his challenge.

I said to the caller,

'Tell him to go fuck himself, I will be there and will show him what fighting is all about, outside of the Karate world I don't fight to rules and regulations, I fight to win and I fight dirty'

The next day I turned up and as a back up I had a few of my Glory Boy mod pals lay in wait, in case I was being set up to receive a gang kicking. I was so hyped up to get it on. The adrenaline had been eating away at me all night and by now I was ready to explode. I was also scared but knew it had to be done. I realised that my Karate instructor was just another form of a school bully and needed to be put in his place. We waited and waited but he didn't show. I went to a phone box and phoned his house. His wife answered and said he was at a student's house giving private lessons. The student was obviously female but to make matters worse she was the girlfriend of one of his other brown belts.

I obtained the phone number and contacted him there.

I said,

'Its Jamie, I'm at the waste ground waiting for you, you coming for it or you to busy shagging Lee's bird?'

He replied,

'Jamie, mate, I don't know what your talking about. I've got no axe to grind with you; in fact I've just put you forward for your Dan grade examination. I will give you all the private tuition you need to get you through'

I said to him,

'You are a sad worthless cunt who gets off on bullying and fucking your students; you are in a position of trust, which you abuse. You are a disgrace to Karate and all that martial arts instructors represent. I wouldn't piss on you if you were on fire'

I then put the phone down. I was so angry and hyped up. It would not have mattered if I were talking to the best fighter in the world. I had lost my passive side to let loose an enraged animal that under normal circumstances remains hidden.

A few days later I had a call from the chief instructor of the association,

'Jamie, I have been informed that you tried it on with your Sensei at the Karate demonstration, forcing him to make contact with you, then you phoned his house challenging him to fight you, what the fuck is going on?'

I said,

'I'm at my mothers house and do not want to discuss this in front of her, so can I come down to your Dojo to explain what had happened?'

He agreed and I went the following Saturday.

When I got there I didn't really expect a friendly reception because it was his word against mine and he was one of their instructors and I was just a student. The chief instructor greeted me and said,

'Jamie, there is no need to explain your side of the story. We have another student from your club who has told us exactly what's been going on and how you were ridiculed and bullied in front of an audience, plus the meeting that your instructor set up to fight you. While you're here you may as well train, we're doing squad training'

I looked around and realised that I was the only person there that was not a black belt, then saw my mate Lee from my old club.

I said,

'Lee, what you doing here?'

He replied,

'I found out that our instructor was shagging my bird, in fact it all came out because you phoned him at her house the night he challenged you, so I fucked her off but had nowhere to train so contacted our chief for another club to train at. He asked why and I told him. That naturally led on to him asking about your situation, so I told him everything that happened. He said that he was going to kick out our instructor from the association but did not want to lose me as a student so allowed me to come down here for squad training'

I was glad that the truth was out and that my instructor was seen for the rat that he was. The training session was the hardest I had ever done because I was now training with 1st, 2nd, 3rd, and 4th Dans who weren't giving anything away. At the end of the 4-hour session the chief instructor awarded me my 1st Dan and the same for Lee. He said we were both overdue for it and realised that we had both been held back. We were to receive our certificates at the fundraising award ceremony later in the year. This was the normal way of doing things so that we would receive our award and have photos done etc, plus the association would make lots of money in ticket sales from friends and family. It also acted as a safety net for the association to stop students getting their black belt then pissing off. I was so proud to be a black belt. It meant so much to me.

I was still very immature because I was still doing all the stupid embarrassing things that new black belts do. For example, on my way to training I would stop at a shop to get a drink, but make sure that my black belt was hanging down below my tracksuit top so that people could see I was a black belt.

Or, even more pathetic, I would take off my Karate suit and wrap my black belt around it, then put it on the rear parcel shelf of my car for the whole world to see. I had an old Singer Gazelle car plus a mod scooter at the time. I was a real sad immature person, but a proud karateka. I loved Karate.

I was a proud black belt

Things at home were getting worse with Karen's drug abuse and my mother's alcoholism. I didn't know how to handle it. If any problems came my way, I would always deal with them physically and enjoy doing so, but when your family are on drink and drugs, it takes a different type of solution and I didn't know how to do it.

I was becoming very emotionally upset and stressed. I was starting to hate the world and was developing into a very bitter person.

I decided to deal with my sister's problem first. I was going to go straight to the people from where she was being encouraged to sniff glue and take all types of drugs.

She was hanging about with a mob of homeless skinheads who hung about London's Trafalgar Square. They had a squat somewhere nearby, a safe house where they did their drugs. I went to the squat firm-handed with the gang and knocked on the door.

The skinheads didn't know me, or who I was, so I said,

'I need somewhere to score, I've got some good gear, which I will share out'

With that the door opened and the biggest, fucking, ugliest, skinhead I had ever seen looked at me and said,

'Who the fuck are you, one of them fucking mod wankers?'

Bang! I hit him so fucking hard he instantly dropped to the floor, and then we all steamed in. We beat them so badly that we thought we killed one of them. He lay there in a coma. We pissed off in a hurry.

Before we left I grabbed one of the skinheads and said,

'Go and phone a fucking ambulance now, get your mate seen to. This visit was because we don't like glue sniffers'

We later found out that the guy in a coma had overdosed prior to us getting there, so we had not caused his condition, but it did scare me. It could have been murder. I justified my actions to myself saying that they were slowly killing my sister, so deserved a good hiding.

The end result was that it changed nothing. I felt a sense of satisfaction punishing the people that were encouraging my sister to do drugs, but nothing changed. Karen carried on doing drugs and they carried on using the squat as a safe house.

My mother was using Karen's drug problem and Peter being in prison as her reason for drinking, Karen was having flashbacks to her childhood, being abused by our biological father, as her reason for doing drugs, she said it blotted the memories from her mind. I personally think that there is a weakness in our family bloodline that attracts us to drugs and alcohol. It's a known fact that drinking

52

alcohol is commonplace for many of the Scottish and Irish and we had both Irish and Scottish blood in our makeup.

I had to make sure that I stayed away from temptation and for some reason my martial arts obsession helped me to do that.

The mod gang had begun to establish its core of main characters, essentially the Glory boys. We adopted the name from a song by a group called 'Secret Affair.' Around this time Paul Weller was sickened by all the violence that went hand in hand with his following, essentially, us. Also at the time the Red, White & Blue Union Jack colours that the Jam used as their trademark, was also being used by the followers of the National Front, British movement, Combat 18 and other racist right wing movements, which we also did not want to be associated with. Weller was now making an appearance on 'Top of the Pops' and wanted to disassociate himself from us. Basically he was making it big time and wanted to lose the thugs around him, quite rightly so. We were trouble. However at the time we were not mature enough to understand that. We just thought that now he had hit the big time, he was shitting on us. We slowly found that we were getting banned from gigs. We were music's equivalent of what football thugs are now. If we couldn't find anyone to fight us, we would scrape the barrel and offer out challenges. The deal was our top man against their top man. If they did not take up the offer, we would kick off with them anyway. If our top man lost the fight, we would kick off as well but if our man won the fight, we would leave it at that.

Dave L was our top man. He was a fighting machine. He was a good boxer and boxed for Repton Boxing Club at the time but was also a good street fighter. He had absolutely no technique or style, but had balls. He would fight absolutely anyone.

One day we were out scavenging and some of the firm had spotted a teddy boy. The chase was on, like a pack of dogs after a fox. The teddy boy had nowhere to run and must have been well worried. I was trailing behind with Dave L, talking about Karate versus Boxing, he said that in his next street fight he would use Boxing only, and to show me how effective it would be for me with my kickboxing training.

When we got to the lad, about to be taken apart by our cowardly mob, I realised that I knew him.

It was my old buddy Pete Chamberlain, who I last saw when he told me to take off my crash helmet so he could punch my fucking head in. It was then that I realised that I also had feelings outside of the gang environment and did not want to see Pete get hurt.

'Leave him, he's my mate' I said.

Pete must have been the most surprised and relieved person on the earth.

That day forged a lifelong friendship between us and we are still the best of pals, over 20 years later. Pete eventually joined up with us, shed his quiff and bought himself a scooter.

Our gang was now established and consisted of myself, Dave L, Ricky and Danny M, Steve B (Crank), Pete C, Danny H, Ian S, Ray W, Darren M and Boris as the main core with many other similar sized groups that were linked to each of us. This formed the Glory boys.

Me (drunk), Ricky M, & Pete C - 1979

I will be releasing a book at a later date (2002/3) based on the exploits of the Glory boys. Within this book I can only include around 1% of our story.

Our main year for violent encounters was 1979 and the rest of this mod era happened throughout that year. When I was around 18.

The Jam were playing Top Of The Pops, and we decided to invite ourselves along to get into the T.V. audience. When we arrived Paul Weller's father, John, saw us and asked us not to kick off or cause any trouble because this was an important day for the group and they were going to be banned from the show, if they attracted any problems. We struck up a deal with him to be good in exchange for guest list entry to the Jam gig at the Rainbow. The deal was set and we played fair. There were a lot of soul boys at the show dancing around with their wedge haircuts giving it the big'un to us but we left it alone.

We were still crazy about the Jam because the words in Weller's songs seemed to represent everything that we were going through as angry teenagers.

As we were banned from more and more places other bands would welcome us, because they realised the impact we could have on their popularity.

We were now following Secret Affair, who were a brilliant mod band with heavy soul influence, and the Purple Hearts, who were kind of a punky mod band. Mainly because they were local bands from Ilford and Romford. The usual fights began to develop because none of us had matured and could see no further than that day. To get round the problem of being refused entry to gigs, as spectators, we decided to get round this by setting up our own bands. It never was meant to be anything serious, more of an offshoot of the punk thing. If they could do it, we could.

I set up a band with Pete Chamberlain, called Untamed Youth; I couldn't really play any instruments too well, so I covered the vocals. We set up a series of gigs all over the country and had the rest of our firm as roadies, managers, and security or just on the guest list. If we couldn't find a position for someone, we would just get the others to set up another bullshit band. It was just a crack, nothing more. A big piss take. Everywhere we went it kicked off. We would see it

Er! any requests?

55

go off in the audience and that was it, we would dive off the stage and steam in.

Whenever bouncers were used at gigs, they got it as well. We operated in sheer numbers and could not be controlled. If any of the bouncers managed to get us on our own, then we would get a good hiding, but in turn we would always go back firm-handed and set the record straight. We did not let anybody get away with anything.

We had an unwritten code of conduct that we adhered to and knew when to kick off and when not to. Also there were some things that we obviously did not condone such as mugging, hurting anyone that was elderly, disabled or a bit slow and so on. If we ever saw any form of victimisation, we were on the case acting as saviour and enforcers of punishment. We certainly had double standards without a doubt.

Untamed Youth 1979
Becontree station

The band was getting successful within our own little world and before the year was out we had released a single. It was absolute crap but must have been worth something to somebody out there because I was recently shown an edition of Record Collector No 182 Oct 1994 and in it our single was valued at £15 but now sells at £50 as a collector's item and is also in the Guinness book of singles.

Crank was one of the more noticeable members of our pack. He was an absolute lunatic. He would do absolutely anything to become the centre of attention. His parents were Greek and owned a chip shop in East Ham but he didn't want to work with them. He would get most of his money doing the fruit machines. He used one of those electronic spark ignition lighters that you would use to light up your gas cooker, to fiddle the fruit machines. He would click it close to the area that gave you credits when you put money in and the electric spark would clock up coin credits for him. He would then play the machine as normal and come away with anything from £50-£200. He never got caught.

He would also do the same to the jukeboxes in the pubs and put on

The Jam and 60s groups all night. He did some really weird things. I can remember him once shitting in a pint glass and topping it up with beer. He then swapped the pint of beer with someone else's drink and stood next to them drinking their beer. It was pretty dark in most clubs so you could not see too clearly and even if you could, not many people inspect their beer every time they pick it up. Anyway, he was standing next to this bloke, who was quite unaware as to what had happened and the guy was about to drink his beer. Just before he took his first sip Crank introduced himself as 'Crank', as he shook his hand, and asked the guy if he had any tissues on him. The guy asked why and Crank said,

'Because I've just had a shit and there was no toilet paper, I had to just use my hand to wipe my arse'

The guy was gutted because he had just shook Crank's hand, but laughs because he thinks he was joking. He wasn't, he had shit all over his hand. The guy then saw the floating turd in his beer and its kicked right off. The thing about Crank was that he had the balls of a lion but the fighting ability of a Barbie doll. Everywhere we went he caused a situation to kick off. He was so fucking funny. You would not believe the stunts that he pulled. Read my book 'The Glory boys' and you will see what a character he was, absolutely unbelievable.

Each member of the Glory boys had their own individual character that shined through and I firmly believe that this group of individuals could never recreate itself again.

Dave L was the pack leader. An out and out fighting machine. No matter whom he fought, they just could not put him away. In one battle at Dagenham Heathway he was stabbed through the back with a pair of garden shears. They have big fucking blades. He survived it OK and kicked fuck out of the rival skinhead leader that stabbed him. On another occasion, at a nightclub in Ilford, a group of army boys tried they're hardest to put him away in the toilet, after it kicked off with one of their mates a little earlier. Dave was badly beaten but the five of them still could not put him away.

Ricky M was a year or two younger than the rest of us, but what a handful. He was only small in height but packed with muscle, and balls of steel, plus he could box. He was not the type of fighter to get involved in grappling.

He was more of a one-punch knockout street fighter, and that he did. He was a great person to have next to you when it kicked off but he did not draw the line anywhere. He would not think twice about replacing the punch with a bottle or iron bar. The last time we were in a situation together was around 5 years ago when 15 guys from our local roughhouse pub decided that they didn't like my reputation as a martial artist. Me, Ricky and 20 pints of Lager took them up on their offer of a fight, they backed down.

Pete Chamberlain had now made his mark within our pack. He was never fight shy and would get stuck in whenever the situation arose. He had a lot of aggression inside him, built up from a rough upbringing, and would unleash it on anyone that crossed him. He had left behind that flash cocky Rock and Roll lad that he was at school and had transformed into an angry young man ready to take on the world. We were at a gig one-day and he was chatting to Billy Idol. Idol was in the band Generation X at the time and was just in the crowd at some venue. He had said something to Peter, which for some reason pissed Crank off. Resulting in Crank offering to kick Idol's arse and that of his two big minders. The King Rocker declined.

We had a thousand and one incidents throughout the mod phase but one that I will never forget was the pasting we got at Ladbrooke Grove, when mobs of skinheads invaded the gig and give us a good seeing to. We were low on numbers and totally unprepared for the attack, but life is such. We were watching a band called the Chords in Achlam Hall when someone gave us the tip-off that there were some skinheads outside that were going to sort us out. This was not something that normally bothered us so we saw the gig out to the last song and then left the building, to have a little fight and go home. I suppose that there were about 30 of us from Barking and surroundings, East London areas, all going home on the same tube, so we felt pretty safe. When we got outside and saw the size of the gang waiting for us we knew that we had bitten off more than we could chew. They outnumbered us on average about ten to one, plus they were tooled up. We couldn't even get back in the building because they shut the doors on us so that the trouble would not end up inside.

We heard a mighty roar and they steamed us.

We had to fight for our lives. They had us pretty much surrounded and cornered so we had no choice but to fight, just like an animal would with no means of escape. It was a horrific battle with people getting bottled, kicked, hit with hammers, cut with blades. We were used to fighting and put up a good fight but the sheer amount of people steaming us took its toll.

I don't know how a death didn't occur. Eventually we managed to break through their attack and made a run for it towards the station. The fighting went on in front gardens, in the roads and ended up on the railway lines of the station. I can remember standing at the stairway of the entrance of the train station with Danny M, holding a broken bottle in each of our hands, just looking at the mob of about 30 blokes running towards us. It was fucking scary but we still found it in us to scream out at them,

'C'mon then if you want some'

They carried on towards us and we left the scene like lightning still being chased by this mob.

We took a right hammering that day. It took some time, but we eventually repaid the ambush and gave plenty of interest in return. Anywhere and everywhere that we came across a Ladbrooke Grove skinhead, we served them up.

My stepfather was still in prison, my mother getting worse with the drink, and Karen was going a stage further with her drug problem. She was stealing, robbing and doing anything she could to feed her addiction. It did not take too long before she and a friend were arrested down Petticoat lane, for mugging and robbing somebody. It was so serious that it went to the Old Bailey but fortunately her pal was a policeman's daughter and things were taken care of. She ended up with a suspended sentence and was ordered to seek medical help for her drug problem.

It's not hard to see why I've ended up with such hate for drugs and those distributing them. Drugs were associated with the mod, skinhead, and punk scene and were a goldmine for the dealers with more serious drugs to infiltrate. They were dealing with a load of immature teenagers who would take anything for nothing and would

59

try anything out just for the crack.

Boris was one of the weakest willed within our crowd and remember him breaking into a chemist and getting caught because he got out of his head whilst on the premises and passed out. The police found him with his pockets full of drugs and stoned out of his head. Crank was also going along that route of always being pilled up. He was a very insecure person who eventually hung himself about five years later. When we eventually disbanded and grew out of the mod thing, he could not adjust and jumped over some banisters with a rope around his neck. We all went to his funeral and that was the last time we were all together as the Glory boys. Sometime after that Boris also died of an overdose. We all went our own ways and tried to integrate into normal life, far removed from the continuous flow of nightlife, gigs, fights, friends, etc. It was now time for all of us teenagers to mature and grow up.

Crank and Boris were dead, Meathead got life for murder, Dave L went to work in Germany but came back just after his brother Terry got shot on the estate. Terry is paralysed for life now. Danny and Ricky M went touring with the East London punk band The Cockney Rejects, which was a natural direction for them to go, being into the West Ham thing. Danny M matured and ended up as an Equal Opps training officer in the Fire Brigade and Ricky M ended up with some heavy stress and mental problems but is ok now, just about. I spent a year talking to him through his letterbox because he was not ready for the big wide world. Danny H set up Centerforce, the pirate radio station with some other characters from the notorious I.C.F., Pete Chamberlain settled down and is still working hard supporting his family. As for my family and me, well there is still a lot more to tell, so read on.

Reaching Maturity

My stepfather had been released from prison again and was now back home and everyone he had ever conned, ripped off, hurt etc came to Dagenham looking for him. The papers did everything but give our house number out for the public to use.

Karen went away to work in a holiday camp but had to come home after robbing the people she worked for. She ended up coming back home, getting pregnant and getting married. Within a matter of months her house had transformed into a drug den and she was now a heroin addict. When her daughter was born it was only a matter of time before she handed her baby over to my mother to raise and she split from her husband. Heroin had now taken control of her life.

I had also been through a nightmare marriage, divorce and had a child, who I was allowed to see once a fortnight. The loss of my son was the only thing on my mind and was becoming an obsession. My life was absolutely devoted to him and when I was denied seeing him for 26 days out of every 28, it tore me apart.

I, more than ever now, wanted to kick off with anyone who gave me a second glance. I was an angry 21-year-old, looking for any excuse to unleash the aggression that I was feeling inside. I was running my own, full contact, Karate club at the time and was going through, what I believe was, the hardest training period in my life. I was finding that more local people in my area were getting to know of me and that I was a karate man so avoided confrontation with me, no matter how far I pushed it. I was turning into a right bastard and also foolishly turning to alcohol. I still have the beer gut as a constant reminder.

I did not have a person in the world I could share my problems with. My mother and Stepfather were worse than ever, on the drink, and Karen was still lost firmly inside her heroin world. I was alone, depressed, bitter, hateful, angry, and did not trust a sole. All I could think about was my son and how badly it hurt not having him around. I wanted so badly to kill his mother and the only thing that I think stopped me, was that, I knew I would end up doing life and my son Jamie would end up in a home. Although I sort of knew, that I was never going to do it. I still studied and researched death of every shape and form, trying to find a way that was foolproof. I was literally

61

going crazy!

I am no different to most other people in this world. Most people have hated someone enough to want to kill them. Luckily time heals and we get over these feelings, but I do not know anyone that can truly say that they have never felt like that, at some stage of their life.

I was drinking so much that I lost sense of reality and was probably a danger to myself. The only time I ever felt sane was when I had my son for the day and I took him out. The very second I dropped him back home I would go crackers again.

I was homeless, jobless, and getting by just on my martial arts club,

I was around 22 years old then and just beginning to face maturity. I think that having lost my home and child really woke me up to the fact that life wasn't a game anymore. However, it was a hard and bitter way to learn it, which I wouldn't wish on anyone.

Karen put me up in her spare room, but it was frightening living there. The place was always full of junkies. She attracted all the low-lifes in and around the Barking area. Heroin addicts have a common interest that draws them together like a magnet. If you are the one with the safe house from which they can 'chase the dragon' or inject that shit, you will have a full house.

I would lock myself away in my room and only come out when they were either gone or asleep.

One night I had been out drinking and collapsed in a drunken state on my bed, forgetting to lock the bedroom door. I awoke in the middle of the night, with flames all around me. They had set light to my bed, with me in it, and were standing around me laughing. I grabbed my stuff and went. As much as I could handle myself, things like this would freak me out. As much as I dreaded it, I went back to my mother's house for a while.

The situation was really strange. I was now a man and my stepfather did not like the idea of anyone challenging his position. He knew that he could no longer bully my mother, or me, because I could punch fuck out of him. Although physically I could have sorted him out, without a problem, I was still secretly shit scared of him. Inside I did not yet feel like a man. He was the only father that I really knew and had felt weaker than him for the 10 years that he had been my stepfather. Every time I heard my mother and him argue I dreaded it getting heavy, because I knew that the time had come for me to intervene.

Money was now not a problem because he was back to his old tricks and literally rolling in it, but his drinking was getting worse by the day. His mouth was getting nastier indoors, and more trouble was finding its way to our house from police, villains and his victims.

Drink and pressure took its toll on my mother' till one day she could take no more. She downed a cocktail of dry Martini and added every pill that she could find in the house. I remember coming home to find the neighbours, in our house, waiting for the ambulance to arrive. She had overdosed in an attempt to end her life. It's a terrible thing to come home to. The hospital did a wonderful job of taking care of her. It also gave her a few days to dry out. It had been a long time since I had seen my mother without alcohol in her system. She was such a beautiful, kind person when she was normal, without drink. The drink would change her for the worse and eventually took her life. Its such a hard thing to deal with, when your close family or friends are killing themselves with alcohol or drugs, right before your very eyes. They cannot conceive the amount of pain and torture they are inflicting on the people that love them. I have felt this pain everyday of my life and still feel it now.

There was no such thing as a happy, loving, traditional, family scenario within which we all lived. It was hell, but a learning environment, from which I matured rapidly.

A little more time had passed and things carried on as usual. My mother was out of hospital but slowly working her way back in with the devils poison, whilst my stepfather carried on with his bad ways of earning money and spending it on booze. Karen was now a registered drug addict and was receiving Methadone, on prescription, for her heroin problem. It was a real shit environment. The people that lived next door to us must have gone through hell over the years. One of the families next door were very understanding and their son Phil would talk to me about music and stuff because they knew I was in a band. They were going to let us play some gigs with their band, who had just released an album, but I just had too many problems at the time to follow it up. It wasn't until years later that I found out that the band were Ultravox, and it was Phil Shears and Chris Cross who formed the band, which was later fronted by Midge Ure. I certainly missed the boat there.

I had so much aggression inside me and found that I had to fight to release the pressure that was building up. One day it was set up, for this skinhead mob, to go and kick off with a pub full of Rock and Roll lads in Bow, East London. I jumped at the chance to go because it was going to be a visit to the Duke of Wellington pub at Bromley By Bow, which was home to teddyboys, rockabillys, and rockers. Many of these were the original lot that turned against me, when I would not turn my back on my sister for being a skinhead.

I did not know any of these skinheads, I just heard about the visit, through the grapevine.

Anyway, we drove up there in an old van and I was in the back with about ten skinheads. One of them farted and the next thing I know it's kicked off. We were all punching fuck out of each other as the van drove to the slaughter. I took a steel toecap to the head and a few punches before we even got there. We had calmed down by the time we arrived but our adrenaline was rushing through our bodies like a speeding train. The plan of action was that one of us would walk in and order a pint. Within seconds the stares and insults would come in our direction so we would leave pretty fast, acting as if we were frightened. It was almost 100% guaranteed that a few of them would come outside to give us a kicking, for daring to drink in their pub. That would then be the signal for the rest of our boys to suddenly appear from nowhere and kick their fucking arses.

Two of my teddyboy mates 1977
Duke of Wellington - East London

I was beginning to feel uneasy in the company of these skinheads and did not want to be stuck in the back of the van with them, so volunteered to be the bait that went in alone to the pub. I actually used to drink in this pub when I was into the rocking scene so knew that there would be a few people in there that would love to kick my arse.

I went into the pub and instead of going to the bar, ordering a drink and waiting for the intimidation to begin, I introduced myself by shouting,

'Who wants to fucking know out of you bunch of wankers,' meaning who wants a fight. I haven't got a clue why I did it?

All I know is that everyone wanted to know.

I got punched through the double doors and was sitting on my arse waiting for the cavalry to arrive. Unbeknown to me, they had fucked off. They had set me up. What I didn't realise was, they had originally planned to get me into the van making out that we were going to kick off with teddyboys, but all they wanted to do was kick off with me or anybody else that would fall for their plan. The tart was their trigger when I was supposed to get the beating, and then get thrown out the van leaving them to move onto their next victim. What they didn't count on was the fact that I could hold my hands up and would not shy away from a fight. I later found out that, when they had steamed me in the back of the van, I laid out their main man, their leader. So, they decided to make me feel uneasy enough to want to volunteer to go in the pub, rather than take a big-time beating in the van. The problem with this was, they fucked off as soon as I entered the pub, so I was on my own. So I was sitting on my arse outside the Duke of Wellington, thanks to some 40-year-old teddyboy, who didn't know I was a black belt. Probably the best lesson of my life but I did not realise it as such until many years had passed.

In philosophy they say that if you are in the woods and a tree falls down, it will make a noise. However if you are not present and a tree falls down, does it still make a noise?

In comparison, with many of the fights I have had, it was known that I was a black belt so this affected the decision as to whether someone wanted to fight me or not. When I entered the Duke of Wellington pub, and offered out everybody, the guy that banged me through the doors did not know I was a black belt. In turn this did not affect his decision, so he just laid me out with one punch.

So the tosser skinheads, who set me up, had deserted me and their wish was about to come true. I was held down by two teddyboys, who were middle aged men, whilst a third one stamped on my face with all the force he could until my nose was almost inverted. My eyes closed up from the 8th or 9th blow, and my lips opened up like pitta breads.

I felt the force of two feet landing full blast with the guidance of gravity, supported by around 16 stone in body weight to my ankle and then a follow up to each wrist. The next thing I remember was being lifted up, like I was going to get the bumps as you would on your birthday, then being thrown over the station overpass, directly onto the railway lines, which must have been 30 ft below.

I woke up in hospital, having been cleaned up and stitched, then allowed home. I think it was around this time that I wrongly began to loose faith in Karate.

I had it in my head that a black belt could wipe out one attacker with ease and multiple attackers would just take a little longer, just like the films. That day made me feel different. On reflection I now know that it was not Karate that lost the fight, it was me. I wore my imaginary black belt with pride and, for some stupid reason, also thought that other people naturally knew and saw that I was a black belt. I was borrowing strength from the concept of the black belt but was becoming weaker in myself as a result. Borrowing strength can sometimes cause weakness. Many, many, black belts do this. They physically take off their black belt when they leave the club, or when they return back to normal street clothes, but they still wear their imaginary black belt, which they can feel and wrongly think that others can see.

Around this time a letter arrived for me from the local council offering me a flat on the notorious Gasgoine Estate in Barking. It was a shit hole within a shit hole, but it was somewhere I could be on my own and shut out the world. I took the flat and within a month it was kitted out like a martial arts training club. I had the heavy bag screwed into the ceiling for my kickboxing workout, speedball in the next room from floor to ceiling, medicine ball, bean bags fixed to the walls, mirrors, and so on. All I had was my training to keep me going. I lived right next to the park so would go running in the middle of the night. As I ran I would imagine people laying in wait for me and at random I would suddenly stop and perform a shadow boxing attack on imaginary people. I was dying for someone to really jump out on me so I could kick their arse. I was totally obsessed with the martial arts and fighting. Not the sporting kind. The real stuff. My old school pals Pete Chamberlain and Bob P were now training with me as well. It was a brilliant time in my life for training.

Karen was getting deeper and deeper into trouble because of the drug thing. Every week she was calling me because someone had ripped her off, hit her or something else. She was hard work.

I was in a very strange situation because I was very anti drug, yet I would have to help my sister out every time she got ripped off. It's very easy on the outside to say,

'There all junkies, let them get on with it, they don't deserve any help'

But when it's your own flesh and blood you do see things differently. Karen was a weak natured person, surrounded by blokes who wouldn't think twice about hitting a girl, just like they would hit a bloke. That's something that I could not let pass me by undealt with. The scumbags that would frequent her flat knew of me and understood that I would pay them a visit if they hurt my sister, but once they had a bit of gear inside them, they didn't give a fuck. Time after time I would have to find someone who had taken liberties with Karen and sort them out.

On one occasion some tosser beat her up in her flat because she didn't hand over her social security book. Eventually after finding it he stubbed out his cigarette on her face, as a warning not to say anything to her brother (Me), otherwise he would come back and inject battery acid into her veins. The correct route to deal with matters like this is the Law, but people within the drug user world, like Karen, do not invite the police into their home unless unavoidable. The only reason I found out what had happened was that I paid Karen a visit and saw the burn mark on her face. As soon as she told me what had happened I was off. The guy that did this to her lived on the other side of the estate in a tower block, which had a camera view entry system, so I had to lay in wait. I sat there for about three hours continuously going over in my mind how I was going to make him suffer. In the end it went from the one punch knockout, right up to tying him to the bumper of my car then driving for a couple of miles. Your imagination plays strange games with you when you have a problem to deal with. Reality time had arrived. 'P' had just been dropped off by someone and had entered the block. I now had to wait until the next resident entered the block so that I could gain entry at the same time. The adrenaline rush was making me feel sick, but supercharged. I was shaking like fuck. I didn't know why because confrontation with this person did not worry me in the slightest, but I couldn't control it.

I was twitching and moving about like a hyperactive kid as I made my way over to the tower block entrance. I was bouncing about waiting for someone to arrive and let me in when suddenly the door opened from the inside. It was 'P'; he had come back downstairs for something and came bowling through the door without a care in the world.

I took him straight down with a double leg sweep, combined with a reverse forearm smash, which all happened within a fraction of a second. He groaned and moaned and looked up at me. As he made eye contact with me I launched a roundhouse kick full blast into the side of his head, which put him out for the count. I ran to my car and off I went. My legs were trembling so badly that I had to pull over and get out and walk. I couldn't operate the pedals properly. Justice had been served, which in the eyes of the law was wrong, but to me felt fucking great. Karate had also worked for me, so I was on a real high. I certainly felt at home with the pre-emptive attacking method of dealing with a situation rather than giving the other person a chance to prepare so that I can be a gentleman and fight by unwritten rules and codes of conduct. Fuck all that. Rightly or wrongly, I felt good and that was all that mattered to me, at that moment in time.

The problems that I became involved with through Karen were on going. Every time I spoke to her something else had happened and someone had taken a liberty with her. It got to the stage that the druggies would expect a visit from me, so would carry weapons, or try to go into hiding, until I forgot about them. After a while they would forget that I was on the warpath for them, which in turn made them think that I had forgot about them. They were so wrong.

You cannot hide for very long in Barking, especially if you are a drug user. They could be found sitting down the social security office on giro day, due to the fact that they were all on personal issue for cashing their giros, then making out they never arrived. They could also be found at the late night chemist, where they would collect various drugs with stolen or altered prescriptions. Plus throughout the day they could be found shoplifting. This was something at which they were very amateur because they eventually got arrested and charged with shoplifting, but in another way also very good at because they knew how the store guards worked and their methods of practice.

Store security guards are under strict instruction that they must work to a method they call A.S.C.E.N.D. (Ascend). This is the qualifying criteria that a shoplifter must fit into before they are stopped for shoplifting, to avoid false arrest. The shop and security companies do not want to be sued for making a false arrest. To ensure this they work under the agreement that if they are not 100% sure that the shoplifter has the goods on their person and has taken them from that particular shop, they must not approach or stop them. So to ensure this, they follow a simple set of procedures, which are Approach, meaning you must see the customer actually approach the goods. Select the goods. Conceal the goods on their person. Exit the premises with the goods. Not offer to pay for goods. Detain them. Throughout this whole procedure the store guard must have 100% continuous observation, so as to ensure that the goods were actually selected from that store and are still on that person when they exit the building.

The shoplifters know how serious it is for a store security guard to make a false arrest, so literally run circles around them. Especially if the shoplifter works in a team. How can a guard have 100% observation on more than one person? The shoplifters in Barking gave the store guards a hell of a time. Another thing they would do was go into a major store and pick up the till receipts that people had discarded. Whatever the item was on the till receipt, they would go and collect from the shelf and just walk out with it. No store guard wanted to stop someone that had a till receipt on them. That wouldn't go down to well in court. Anyway, I'm drifting away from the point. If it was time for someone to face the music, because they had hurt my sister, they could run but could not hide for long.

Steve was a longhaired, denim and leather type, who had broke into Karen's flat and robbed her electric meter. There was about £70 in it.

We knew how much he took because my sister had a key made that fitted the lock and would continuously borrow from the meter and pay it back, before the meter man called to empty it. It goes without saying that Steve was a junkie like all the rest of them. I decided to pay him a visit, to collect my sister's money and discipline him for the break-in. Like most of the others he lived in a flat in a tower block on the estate. I knew he wouldn't open the door to me so I took my sister to knock on the door, so I could gain entry. She knocked and called through the letterbox to him. He looked through the spy hole and saw it was her on her own and began to unbolt the locks. All junkie houses had loads of locks on their doors. As the door creaked open I launched a powerful spinning back kick to the centre of the door, which hit Steve sending him flying back with force. I steamed in and launched an array of martial art techniques on him. Somehow we ended up going through his living room door, in a kind of grapple. I looked around me and saw a room full of rocker type blokes. Fuck it, I thought, I was now going to get the pasting of my life. I decided that if I was going to take a beating I might as well go down fighting. I screamed out,

'Who else wants to fucking know? You, you, or you?'

As I pointed my finger at each of them. Not one of them said a fucking word to me. I got out of there pretty lively.

I think that, because they saw what a mess I made of Steve's face, none of them were going to chance taking me on and possibly losing. In reality, if they had steamed me I would have been seriously hurt. I think the lesson of the day was **'Look before you leap'**

The problem still existed, Karen still had not got her £70 back, so I had to take matters further.

I had a pal deliver a message to the flat saying that if I did not receive my sisters money the next day, I would be back to do Steve badly.

That night the money was put through her letterbox, with a bit of gear just as a sweetener to calm things down. A happy ending.

Her problems literally went into the hundreds. I could write a volume of books just on that area of my life.

My personal training was becoming more and more geared up for reality. I was training hard in full contact, which comprised of good quality kicks, but shit hand techniques. I was pretty good with bare knuckles but clumsy with the boxing gloves on, as were most Karate guys trying to do, full contact. The good thing, though was, I now had an excuse to do plenty of bag work in a different way. When doing traditional training in Karate, bag work was never part of the training. It was more focused on being able to kick whilst keeping balance and looking good. Also you were training to punch through a wooden board or break a breezeblock. It all looked very impressive but not much use to me.

When someone sinks a right cross into the side of your jaw, it matters not how many boards you can break or what colour belt you wear.

I was specialising on my spinning back kick, which became my most powerful weapon. I drilled this technique thousands of times and used it about 50-60 times in real situations. It was the last thing anyone expected from a tubby guy, which allowed it to work beautifully.

Sometimes I did come unstuck though.

One time I had to kick a door in for a pal who had lost his keys. Instead of trying to push the door in with a forward heel kick, as most sensible people would do, I had to show off and do this flashy spinning back kick. What I didn't realise though was that the old council door was hollow inside and my foot went through both layers, coming out at the other side. I was hanging upside down with my foot firmly wedged into the door. The neighbours came out and just looked at me hanging there. Their children were giggling and laughing at me. I felt a right prick. My mate took ages to prise me free. I was so stupid at times.

Other times I have launched the kick in a street fight and I've either slipped because of the shoes I was wearing, or I've missed the target because they moved. Unfortunately if you practice with a static heavy bag for a long time you are not so prepared if your live target moves. That improved with time, as did the target area. I always kicked to the gut, which would effect the desired result of knocking the wind out of them, whilst also knocking them on their arse. However some would get back up from the floor and come back to kick off again. As fights became more serious and I also accumulated more misses, due to moving targets, I changed the kick to the pelvic bone rather than the gut. This kick would cause so much instant pain that they would just drop to the floor. No getting back up from that one and it actually

71

caused some of the recipients to piss themselves.

It's terrible to think back on some of the things I have done, to some people, but in all honesty, the people that I have seriously hurt deserved it, and certainly would have done much worse to me if given the opportunity.

One scumbag I remember was 'AZ', he was very cocky and confident with himself. Always offering people out, then taking their cigarettes and money from them if they showed a weakness.

We knew of each other but never crossed paths. We moved in different circles. His was the druggie world.

One day I was on my way over to see Karen and as I drove down the one-way system I caught a glance of this guy wearing my tracksuit. I knew that it was not just a similar one because it had my own personalised name embroidered into it, along with my club badge, pretty sad eh! I did not see who the person was because the wearer had the hood up to hide his face.

I was not able to stop because of the way the traffic was flowing, so drove direct to Karen's. It would not be the first time that she had sold something of mine, to feed her habit, and many of my belongings were still in her flat from when I stayed there.

I knocked on her door and her neighbour answered.

'Jamie, thank god you're here, your sister has just been beaten up and robbed'

I hurried into the living room, where, my sister was huddled up and crying.

'I want you to kill that fucking bastard, he slapped me up then took my money and my wedding ring'

After calming her down she told me that it was 'AZ' that had hit her because she had accepted some drugs from him on credit and later realised that it was just baking powder and not heroin, so refused to pay. She did not have the money that he demanded, so he got heavy. Her wedding ring would not come off of her finger so he prised it off with a screwdriver. Her finger was in a real state. Although my sister was divorced, she still wore her ring.

I asked about my tracksuit and she said that he stole it along with some other stuff of mine and attacked her with martial arts type moves, whilst taking the piss out of me. When I say me, he was taking the piss out of the person that owned the stuff, thinking that it was her boyfriend's, saying how her kung fu boyfriend was not there to protect her. Apparently, 'AZ' knew of me but did not know that I had a heroin addict sister. Not the sort of information I offered out too freely. So in turn did not know that he had fucked with me by hurting my sister.

It took me a few hours to find out where he lived and decided to give him a visit. It was going to be a nice early wake up call around 5am.

That's the way I worked, if I couldn't find someone on the streets. I would boot in their front door and serve them up badly before they even had the chance to wipe the sleep from their eyes. Like I said before, there is a way to do everybody and this is one of the methods that I liked to serve up to someone, who has been very out of order. If they had family, kids or a partner in the house then I would not even consider this option. I'm not a callous thug or a villain, but I will not let a personal injustice go undealt with. Most of the druggies were used to their door going in at unusual times of the day or night, but were more interested in flushing the drugs down the toilet, rather than worrying about a kicking coming their way. I took Bob P with me to drive, because I knew that I would have trouble driving away from the scene, because of the way adrenaline affects me. It makes me shake like fuck, making driving a bit of a problem. We approached his flat and it was quite obvious that he was not asleep because he had music playing. It was very common for druggies to sit up all night and sleep through the day. I composed myself and gave the door the hardest kick I could. I kicked it in at the hinge end, rather than the lock end. Most junkies had at least two, 5 lever, locks on their door with well placed dead bolts inside, nicely positioned to stop the police gaining entry too quickly. Not many of them secured the hinge end though, so this was the side I would attack. It took a couple of kicks and part of the doorframe but eventually it went through. 'AZ' was hiding in the meter cupboard. I pulled the door open and he stood there shitting himself. He didn't yet know why I had paid the visit. As he started stuttering and panicking, trying to ask what was wrong, I noticed that he was wearing my sister's wedding ring. I let go with one blast of an uppercut, with my left, which dropped him like a dead weight. I then dragged his lifeless body, into the living room, by his hair and then attempted to take the ring off his finger. Whilst I was doing this, Bob discreetly dealt with a nosy neighbour who was awoken by the door going in. I don't know exactly what Bob said but I think that they mistook him for a copper. He stood 6ft 2 inches and in his security outfit, at 5am in the morning, with a sleepy head, you could easily mistake him for the police. He obviously never said or made out to be a police officer, because he knew he could get nicked for that, but he certainly played the part if someone approached him.

'AZ' was still unconscious on the floor as I tried to get the ring off his finger. It was jammed on and wasn't budging. I went into the kitchen and got a bread knife and started cutting away at his finger. It was his little finger. I was not trying to cut it off, I just wanted to make a mess of it, as he did my sisters, I also wanted him to wake up with me sawing at his finger, to frighten the fuck out of him. Bob came in and saw me sawing away and freaked out. Just then 'AZ' regained consciousness, saw what I was doing to him and freaked out as well. He jumped up like someone had put a firecracker up his arse. I stuck the nut (head butt) in a couple of times to shut him up but he was panicking like crazy. He could not pronounce his words properly because, I had broken his jaw in 2 places, plus he had bitten into his own tongue when I gave him the uppercut, resulting in a swelling. I found this out at a later date. The only way I could shut him up was to stick the bread knife to his throat, as if I was going to stab him. This shut him up. I said,

'The girl you beat up was my sister. This is her fucking ring and somewhere in this flat you have my tracksuit. I will come back and visit you every single day, wherever you are if the gear is not returned. Each time I will break one of your fingers, do you fucking understand me?'

He was moving his head up and down like a nodding dog and pointing over to the corner of the room. My tracksuit was in a heap on the floor.

I went over and picked up my tracksuit and said,

'Is there anything else of mine here that you've got?'

He waved his hand left and right, which was the body language for no, and painfully tried to turn his head at the same time to give the same reply.

I then left and on my way out I said,
'If that ring is not returned today I will be back to cut your finger off,' then walked out.

The noise from his stereo disguised all that was going on in the flat. Can you imagine what it was like living on this estate? Music blaring at 5am and doors being kicked in. It was commonplace and neighbours just did not want to get involved. It was bad enough for them having to live next door to scum, let alone getting involved in these types of situations. Truth be known, they probably loved seeing these scumbags getting sorted out.

Later that evening I had a visit from the police who were asking questions regarding my visit to 'AZ' earlier that morning. Apparently he went to hospital with a broken jaw and amputated finger, claiming that a guy trying to steal his ring had hacked off his finger. The nurses called the police, who spoke to him and in turn paid me a visit. I told them that I had tried to get the ring off, but did not cut his finger off and that at the time he claimed to have been mugged, I was privately teaching a student of mine, which I was. They said that they would contact me again if they needed to speak to me. They were well aware of 'AZ', his life of drug dealing and criminal activities and didn't give me a hard time at all. I'm sure they were secretly smiling inside.

Karen's ring was returned anonymously in an envelope through her letterbox. It had dried blood and bits of skin stuck to it. 'AZ' disappeared from Barking after that and I never heard anymore from the police. I don't know whether 'AZ' cut his little finger off as a mark of dishonour, as the Japanese used to if they were disgraced. Or if someone else hacked it off whilst mugging him. All I know is that for years after that the story still went round, about how I supposedly cut his finger off. Karen even claims that the ring was still attached to the finger when she opened the envelope.

Who knows? All I do know is that 'AZ' somehow lost a finger and for some unknown reason it was never stitched back on.

The problems with Karen just went on and on and still do right to this very day. But what can you do? Family are family.

Even now as I write this book in July 1998, I have received a call from Karen telling me that her boyfriend has been stabbed and has just been flown to hospital by air ambulance. She is still living on the

75

same estate.

After the episode with 'AZ' I tried to get on with my life and try and stay out of Karen's problems. Don't get me wrong! I would never back away if something serious was happening to her, I just needed a break. I could feel her problems literally draining my energy from me.

I must have upset a few people in my time but none as much as the person that put a contract out on me. Sounds a bit 'gangsterish' but that's exactly what happened.

I was going through a period of receiving death threats on my telephone, from anonymous callers, having my car windows smashed and silly shit like that. Probably from someone that I had beaten up, or pissed off at sometime, who didn't have the balls to confront me face to face. I never took these kind of threats seriously because not many people send you an advance warning if they really want to hurt you, they just do it. I was always getting messages that a local thug, 'LJ', wanted to cut me up, because, I had slapped his brother for breaking into my mums house.

'LJ' was hard within his little circle of Barking thugs, who would prey on the weak and gain money, fags, and anything else they could get, through inflicting their reign of fear. He meant 'shit nothing' to me, he was just a local wanker trying to make a name for himself, who, felt put out because I chinned his brother. I had just finished re-spraying my car, which I bought to do up as a collector's car. It was an old Morris Traveller, which was nothing special to look at but in its condition, was worth a grand. I lived in a tower block so did not have a front drive, or garage, to park it in. One night a brick was put through the windscreen. That's the sort of thing that tossers do when they haven't got the balls to fight you, they sneak up whilst nobody is around and attack your property. It was stupid immature kids stuff, but still caused me lots of aggravation. I would have much preferred a fight where I won, lost or drew and it was all over, but the car stuff was just fucking irritating. I couldn't prove he did it but sometimes you just know. I had the window replaced and the following week the tyres got slashed. There was nowhere that I could plot up and hide, to lay in wait for the next visit, so decided that I would get an air rifle and shoot the next visitor who decided to vandalise my car. I couldn't do it from my flat because he would know which window I would have to look out of, so, I hid in the rubbish chute of the facing block. I did this for two nights until I struck lucky. It was around 3am and the area deserted. He approached my car and looked up to my window, to see if I was looking out. Nobody was there so he began to drag a sharp object across the paintwork of the door. I shot him in the fucking arse and made him scream, like a squealing pig. I came out from behind the rubbish chute doors and took a second aim, which encouraged him to take off at the speed of light. I don't know if it was 'LJ', or one of his thugs, because I could not get a clear view of his face. I do know that it was connected, because, the next day a message was relayed to me that I was going to get cut up that night, when I went for my run over the park.

I didn't go to the park that night, because although I didn't take the threat seriously, I just thought it was a trick to get me away from my flat. I just forgot about it. Across the road from my flat was a pub, which 'LJ' used to meet up with his thugs. They could see my window from the pub and would know when I was in and when I went out. I thought that they were expecting me to go to the park, to see if they were going to carry out their threat, then, whilst I was waiting in the park, like an idiot, they were going to break into my flat and rob it. That was their game. Later that evening the police sirens were going off over by the pub and an ambulance arrived. I thought it was the usual end of night pub brawl. The next day I found out a man in the pub had been stabbed to death and that 'LJ' had been arrested for it. He was held on remand for a year and found not guilty. That's the last I ever heard of 'LJ.' While 'LJ' was awaiting trial I was still getting the death threats and other shit happening, like a wreath delivered to my door, shit put through my letterbox, and similar stuff which I believe was unrelated to the 'LJ' thing.

One day I was going into Barking, to get a bean burger, then to visit my mum. It was a Sunday and the town centre was pretty dead. I asked Bob P to drop me off at the Burger shop, across the road from the station. Bob drove me to the shop and off he went. I bought my burger, left the shop and was planning on walking to my sister's flat on the estate, to see how she was. The next thing I knew three blokes dragged me from behind, into the alley next to the burger bar, and I literally fought for my life. They tried to kick fuck out of me. I swear to this day, that if I did not have the practical knowledge that I had at the time from the martial arts, I would very likely be dead now, or be very seriously disfigured and disabled.

I was dragged backwards by one of the guys, who used a rag or piece of clothing to strangle me from behind. He had no problem in pulling me backward off balance. The other two grabbed my arms and used pure force to run me in a backward direction. I wasn't in possession of the defences against rear strangle attacks at that time, so ended up on the floor with them sticking the boot in. I managed to grab one of the legs and did a roundhouse type leg kick to his supporting leg, this caught him nicely in the side knee joint, putting him out of the game. They were not fighters, or professionals of any kind because they would have finished me off. They were still managing to give me a brutal beating, but no more than any one else I have fought before. I now had the guy, who was strangling me, on the floor, with me laying with my back on his chest whilst he was still trying to strangle fuck out of me. I just kept elbowing him until I caught him with a nice one. I don't know where I made contact but I felt the tightness of the rag across my throat loosen. It may even have been a badly placed kick from the guy who was still standing and sticking the boot in, he may have caught his own man in the head. It was now just the two of us, because one was temporarily out for the count and the other was crying with pain from a dislodged knee.

The tosser then pulled out an aluminium comb, which had some teeth missing, and slashed it across my face. My immediate reaction was to cover my eyes with my forearm, just like you do if somebody throws a ball towards your face. It was not a well-practised martial arts block, which I had spent years practising just for a rainy day like this. It was just a simple reactionary face cover-up. The weapon was slashed across me with so much intent and force that it pierced my inner forearm and ripped a 5 inch gash out of my arm, causing tissue and vein damage. I didn't really feel much pain because my adrenaline was acting like a pain barrier. I returned a forceful front kick to his pelvic bone and he went down.

I then grabbed hold of the leg of the guy whose knee I had just damaged and twisted it as hard as I could shouting,

'Who the fuck put you up to this?'

He was sobbing with pain and fear and cried back,

'This geezer in the pub offered us a lot of money if we hurt you badly, the more we crippled you the more we would get' I gave his leg another hard yank,

'What's his fucking name, who the fuck is he?', I screamed.

'I don't know, honestly I don't know, all I do know is that we had to meet him at the pub tonight to get our money'

'What fucking pub' I said,

'The Hope on Gasgoine estate at 6 o'clock, that's all I know'

By this time, the one that was strangling me was coming round and the guy I kicked was attempting to get up to make his way towards me.

The blood was leaving my arm pretty rapidly and it seemed like a bloodbath. In reality it probably wasn't a lot of blood at all, if you have ever been in a real blood drawing encounter you will know that a tablespoon of blood is enough to, literally, cover the walls and ceiling of a room. Blood goes a hell of a long way.

I heard a car pull up at the end of the alley, which was probably unrelated to this situation, but I didn't want to hang about to find that it was more help arriving for them to finish off my beating, so off I ran. My blood was dripping everywhere.

I made it over to a friend's house on the estate who took me to the hospital, where I got stitched up.

That night I went to the pub with a few pals. I wasn't in any fit state to fight anyone, but it didn't stop me from looking around at everyone in the pub. I saw everyone but still saw no one. I did not have a clue who I was looking for. I was up for it though regardless and they knew it!

Nothing happened that night and to this day I still haven't got a clue as to why I was set upon. I learnt though from that day that awareness of danger can help you avoid ambush attacks, like that one, and made sure that I was more switched from then on.

My Stepfather Peter brought just as much trouble to our home, due to his villainous ways, like the time I came home and was greeted with the sight of a shotgun being held to my mums head, by Scottish John and his two gang members.

Peter would set up jobs like, Milk Dairy robberies, Jewellers and similar places, where it was guaranteed to reward the gang with lots of cash, or jewellery, which were very tradable. He would find out all the information, work out the plan of action and then sell the job on to John and his boys, for a quarter of the takings. Again, I want to remind you, it was not something that I condoned, but it happened and I had no choice, being in the family, as to the amount of exposure that I had to their dealings.

Peter had set it up for a Jeweller to be robbed in Hatton Garden, after a tip off he had received from a girl, who had been sacked from the shop. John and the boys went in with shooters, stuck the place up and cleared it out of all they could get into the mail sack, within the time they had. The plan then was that they would drive to a certain point and throw the bag to Peter, making it easier for them to split up and disappear, via, motorbike, tube and bus, all in different directions.

As planned, the sack was thrown to Peter and they disappeared. The arrangement was then to meet Peter later that evening, to cut up the parcel and each do their own thing with the quarter they each had. Peter had different plans, he decided to knock them for the lot, tell them that the police were on his tail and that he had to dump it in the Thames. Hoping that they would fall for this story and then make plans to recover it from the Thames. Peter knew this would not be practical and in the end they would forget about it. Unfortunately for Peter he went out drinking and as he became more taken over by the alcohol, he began bragging about how he had mugged off John and his boys, for the sack of gold. News got back to John and he paid a visit to my mother's to lay in wait for Peter when he arrived home. My mother was a hardened Scot, who had lived through the rougher side of life and was not too intimidated by these wankers. She knew that they were not really going to hurt her, that was not their way. Although they robbed places using firearms. They never hurt anyone and were not tough guys. They just wanted their money. My immediate concern was to get them away from my mother. Although I knew that they would not intentionally hurt her, I still didn't want her to be involved, just in case something did go wrong. If Peter had returned home to this, with a few beers in him, he would kick off with them. That was something about him. He would not back down to anyone, regardless of the consequences.

I told them that I knew where the sack of gold was and would take them, without Peter knowing, just so that this situation could be cleared up without any ugliness. They trusted my word and came with me to collect the bag. My mother was very wise and knew that I was taking them to a car-parking garage that Peter used to store his stolen gear, however this time it was empty. As soon as we left she phoned Peter, at the pub, and by the time I had arrived at the garage, with John and his two goons, Peter had arranged for a welcome party to meet them.

John and his boys were vastly outnumbered and tried to talk their way out of the situation, of threatening my mother with the shotgun. Peter took the gun from John's car and put it to his head and pulled the trigger. It was empty. Peter then rammed the butt into his jaw and eye, splitting his face open. John was begging for forgiveness. Peter then told the other two that they were going to get their legs broken, for the visit that they made to our house, but he would let them off if they each broke one of John's legs. The slimy bastards both steamed into John and kicked fuck out of him, trying they're hardest to bust his legs.

Peter stopped them and explained to them that it was John who had arranged with him to make out to rob them of the gold, so that John could con his own two boys of their cut of the gear. But John thought that Peter had also robbed him of his share so decided to pay my mother a visit. It was a bit complicated, but basically, every one was trying to rip each other off, the thing that they did wrong was paying my mother a home visit. That was not acceptable in the world of villainy, so John had to be punished. Peter handed over some of the gold and told John and his boys to work it out between themselves who got what. Later on I saw what Peter had kept back for himself. He was sitting on a gold mine, which he later converted to cash, was throwing money around for quite some time. Within the gold was a double diamond twist platinum ring, worth around £600, which he kept to use for the 'Tweedle.' This was the name given to another of his scams. Peter had a dozen replicas made of the ring at £12 each. It was the oldest trick in the book and his first step, years earlier, into the con game, or 'corner game' as its known in the trade. He would offer the ring for sale to anyone that had £400 to part with. He would let them nominate any jeweller of their choice for the ring to be valued. The real ring would always be valued at around the £600 mark. Peter would then easily sell the ring to the buyer for anything between £400 to £500. Obviously he would switch the ring with the fake one before he handed it over. The buyer would go away thinking they had a great bargain, but in reality only had a £12 ring. Peter easily did one of these a week. He carried on with the 'Tweedle' for a couple of months until he lost the real ring one-day. His working partner at the time, Hughey, set Peter up with a fake buyer who managed to do the switch on Peter in the Jewellers, before they did the transaction. Peter was not aware of it until the next time he tried the 'Tweedle.'

His partner gave the game away when he went to the same place that Peter went to have his replicas made up.

It's very true that there is no honour amongst thieves.

In the meantime, whilst all this was going on, Scottish John called Peter to ask if they could work together again. Peter told him to fuck off.

So John offered Peter a bigger cut of any takings. He carried on setting up jobs for them, for about another year, but eventually they fell out again after Peter tried to rip them off once more. John and his boys set up their own robbery on a bookie and got caught with the money and guns. They are now all serving long prison sentences.

Although it may not seem immediately obvious as to why I have included all these stories, which, on the surface may seem irrelevant. My exposure to this secret underworld is something that not many people get a chance, even to glimpse of. It all helped me later on in life; I looked at areas of security with a much wider database of knowledge, on how criminals work and the devious tricks that they get up to. Coming from a family that moved in circles connected with drugs, crime and violence you do tend to be a little more aware and streetwise when someone is trying to pull a stroke on you.

I am not trying to glorify criminal activities in any way, but feel its only right to talk about some of the pitfalls, so you can see how that has also affected my life and has given me different experiences.

There was many an occasion when things went wrong and trouble found its way to our door, like the time when this guy knocked on my mum's door saying that he had reversed into the side of the Rover car around the corner, which he was told was ours. Peter saw this as an opportunity to get some money out of the guy, for the damage, and still have a claim being as he had full comprehensive insurance. He went round to the car, to inspect the damage, and was jumped by three blokes and dragged off in a van. It turned out that Peter had conned a publican out of £1,000 for some bottles of alcohol, which obviously never materialised. But somehow he had suspected that Peter was up to something dodgy and had a mate plotted up nearby, who managed to get the registration of Peter's Rover. With this information and a good contact at the car registration centre, they managed to get our home address and put together this plan to get Peter out of the house, to try and get their money back. Peter was taken to their pub, which was situated in the East end, and tied up to a chair in their cellar. They began slapping him up and demanding the return of their money, but, as I said before, Peter would die before he gave any money back. The publican then telephoned my mother and said if the money were not returned, by the end of the afternoon, Peter would get kneecapped. Meaning, they would smash out his kneecaps with a baseball bat, or even shoot them out.

In the meantime the Police had arrived at the door. One of the neighbours had called them to say that some guys had beaten up Peter and took him away in a van. My mother denied all knowledge of it saying that Peter had popped out for a while, that's the way Peter would have wanted it. The Police were pretty pissed off because their time had been wasted. They knew Peter and wanted to find something on him to put him away, they also knew that my mum was lying, because his car was still outside.

As soon as they left I rounded up Peter's money. He never put anything in the bank because he could never justify where it came from, also he spent it too quickly. He must have had about £3,000 in his wallet. He had two wallets. An empty 'Dummy' one, which he carried from habit, from his days on the run. If he ever got captured, the police would take the couple of hundred quid that he had in his 'Dummy' wallet, before taking Peter in, but Peter would also offer them another grand each if they would let him go. Believe me, this happened quite a lot. There were many crooked coppers about at the time. All they had to do was let him go and someone like my mother would turn up at an arranged time and pay them off. Another trick of his was, as soon as he conned someone out of money, he would make his way to the nearest post office and send all the money off to his own address, via recorded delivery. So even if the buyer or the police captured him, he wouldn't have any money on him. This happened a couple of times and they could never work out where the money had gone. The money would always arrive home the next day courtesy of the post office.

Anyway, back to the story. When the Publican kidnapped Peter all they would have found on him would be his dummy wallet containing, at the most, £200.

I phoned the pub and arranged to return the £1,000 on condition that Peter was unharmed and let go once I arrived and on the understanding that Peter did not know that I was there, returning the money. I said it was out of my own savings and I knew nothing about what had gone on.

This was all agreed and off I went.

I stopped at the bank, to change the money into new notes, so that my story looked genuine and arrived at the pub about an hour later.

I walked into the bar and it seemed as if the whole pub was full of heavies. It may have been just your average local pub, with a lot of regulars in, it but to me it felt like every set of eyes in the room were on me. I was shitting it!

The Publican was sitting and waiting at the bar for me. I introduced myself and showed him the money. He gave a nod to another bloke at the end of the bar, who disappeared. The publican told me to look at the rear window for a moment. A short while later I saw Peter pass by and cross over the road and off he went. Peter was still totally unaware that I was upstairs handing over £1,000 of his money.

Peter was safe and I handed over the envelope, which contained the money. The publican flicked through it and gave me the nod that it was OK to go.

As I approached the door two lumps blocked my exit and one pointed his finger at me rather aggressively,

'Tell your fucking father that if he ever sets his foot in the East End

again, we are gonna hurt him badly'
I didn't say or do anything. I just stood there lining them up inside my head and the second it looked as if I was about to get served up, or anyone else closed in on me, I was going to have to push my fingertips deep into their eye sockets and run like fuck. I was trying so hard to control my trembling legs and just wanted to be a million miles away from there. I really thought I was going to get a beating, but just hung on for that few extra seconds hoping it would go away. It is so easy to forget how it feels to be scared, when you are a big fish in your own little world, but when you suddenly find yourself in someone else's pond, you feel like a lamb going to slaughter. Your fighting capabilities are no different; except for the fact that you are shaking like fuck and doing you're hardest to disguise it.
'Let him go', said the publican.
It was like the fucking cavalry had arrived. The two lumps moved away from the door and I walked out. It must have made the two lumps wonder why I didn't panic, or beg not to be hurt. Little did they know? As I got round the corner my legs practically gave way on me, where the adrenal rush had not been put to use and I was now getting all the after affects. Shaking, sweating, short breaths etc, I was in a right state. It took me about a minute to compose myself and then I went home.
Its funny how in all my martial arts studies nobody ever taught me about how it feels, when you think that danger is going to rear its ugly head. Any learning, knowledge or understanding that I have accumulated is through real experiences, in the real world, where I found out for myself how things really feel, rather than how I thought it would be.
Anyway I got back home and Peter was going mental at his money being given back, but also told me how stupid I was to go to the pub. He was secretly grateful but was not able to express it in his frustration about the money changing hands.
What pissed him off even more was the fact that he had about £250 in his dummy wallet, which they also took, so they were £250 in profit. He decided to leave it until they forgot he even existed.
'There's a way to do everybody', he said.
Six months later they were robbed at gunpoint of £15,000. The only

description that the police had to go on was of 3 guys, one with a Scottish accent.

You may wonder why a book about me has so much stuff in it about other people. Well it's because of the direct and even indirect influences that other people have had on me. My stepfather, Peter, had a massive influence on me and my understanding of street life and villainy. It helped me to stay away from villainy myself, which could have been an easy temptation but it was not a path I chose. It's a bit like any child who does not want to be the lawyer, teacher, policeman etc like their. You kind of get sick of it being around you continuously, as you grow up, and it naturally encourages you to choose a different profession that excites you rather than going into a field that you have grown up with.

As time passed I disassociated myself as much as possible with Peter's criminal activities and he kind of slowed down a bit due to ill health. The Old Holborn roll ups and bottles of Whisky took their toll on his insides and he was gradually loosing his mobility.

He finally gave up his criminal lifestyle to become a lorry driver, but even then he found a fiddle. He was working for a bed manufacturer in Canning Town, delivering beds to shops.

He had a deal with the guys that loaded the vans to put on extra beds, which he would sell to second-hand shops and the like. His wages came to around £200 per week but the fiddle gained him an extra £600 per week. It was a safer way for him to earn a crooked income. Needless to say it came on top for him after about a year, because, he was driving around in a more expensive car than his own boss. He had the option to leave by choice or explain his lifestyle to the police. The reason for telling you this is that from there he gained employment for a haulage company in Barking who, later, also employed me to work in a factory that they had making dog food. So we ended up working very closely with each other and began building up a better relationship.

The boss of both the haulage company and the dog food factory was an old guy, called Tom Griffiths. He was one of the old school and loved having Peter working for him. He was always asking him to tell him the stories of stuff he had been up to over the years. Tom was very good to Peter and in turn was good to me. I think this was the first time in Peter's life that he had worked without fiddling. Peter respected Tom and enjoyed working for him. This evened out most of the in-house family problems between Peter and my mother and I remember this time as being the last time Peter ever laid a hand on me. We had fallen out over something and he had grabbed me by the throat and bashed me up against the wall. I did not retaliate; it was not my way. Although he was not my biological father, he was the only father I had ever really known and in turn would not harm him. I felt much different to this as a kid but I had not matured. I think it was on this occasion that Peter realised that he was growing weaker and that he was not going to be able to push me around anymore. This also in turn stopped him being abusive towards my mother. I was around 22 then, was now the man of the house and my opinion on matters was sought when there was a problem.

Peter did not want to accept the fact that he was growing weak and old due to his tobacco and alcohol abuse. He also knew that as he grew older and weaker, he would have to rely on other people and slowly lose his independence. But one thing that he could not ignore was that he could no longer control the family with his reign of fear he inflicted upon us. His power and control over us had gone.

It was only a matter of time before my mother left him.

He could no longer chase and beat her; he could no longer hurt her with psychological abuse. It was now time for her to get a life for herself.

My mother lived in temporary accommodation for a while until the council eventually gave her a flat of her own; this sadly was situated on the dreaded Gasgoine estate.

We were now at the stage where my mother, sister and myself each had our own high-rise flats on the terrible estate. Peter then gave up the house he had and took on a flat on the estate, because he could no longer climb stairs. Then the fun really began.

Karen was at the worst possible stage with her habit and was getting nicked every other week for everything from social security frauds, to theft of doctor's prescriptions. She was up to every criminal activity that you could think of just to feed her habit.

It was now time for the family to pull together to try and help her sort her life out before she either got locked up or died. She was saved from prison so many times, because she had a little baby, but it was only a matter of time before her daughter was taken from her and put into care. To avoid this my mother took care and control of my niece around 1983 so that, no matter what happened to Karen, her child was safe. I was also there as a back-up, to become her legal guardian, if anything was to happen to my mother, or the social services raised the question about my mother being an alcoholic. Karen was happy with the arrangement and it was set up without the need for court involvement. The next stage was for us to persuade her to go into a rehabilitation unit in order to join the drug withdrawal programme. It took some time to get this into operation but we eventually got her in there.

She was accepted into Meta house at Bournemouth, Dorset, which was well out of range of her local drug contacts, and she did well on the programme. A couple of months passed and she was pretty clean and looking healthy when one day we received a call from Meta house saying that Karen had left the project in the middle of the night. The thing with the drug rehab scheme was that it had to be done voluntary; you could not force someone to stay against his or her will. Karen made her way home and avoided us for a day or two but we eventually found her, drugged up and back to her old ways.

Peter decided that he was going to pay a visit to Richard, who was supplying Karen with heroin, to try and stop her before it went too far. In the back of his mind he knew that there is no way to stop a drug user, unless they want to stop, but it was all he could think to do. Peter hobbled out to his car and sat there to regain his breath and then made his way to Richard's house. He pulled up outside the house and pressed his hooter until Richard came out to the car. Peter began an angry barrage of insults and threats, telling Richard if he didn't stop supplying Karen with drugs that he would sort him out. Richard pulled Peter from the car and knocked him to the floor. Peter was too weak to do anything but lay there. He eventually got himself up and back into the car and made it back home. My mother phoned me and told me what had happened and I told her not too worry about it; I would take care of it.

I drove round to Richards house and kicked his door. I could not put it through because of the dead bolts on the hinges. Richard eventually came to the door holding a syringe.

'If you come near me I will stick this needle in you and you will die in seconds' Yeah right! As if I was just going to walk away shaking! I launched a front kick to his chest with the intention of knocking him back, so that he would drop the syringe and I could then give him a beating. I missed slightly and caught him in the throat. It was a lovely one that I would never have managed if I had tried it. He dropped the syringe as he pulled his hands to his throat. He was coughing as if he was choking to death but I think he was putting it on to escape a beating. I then treated him to a right cross, connecting my bony knuckles to the tip of his big pointy chin, sending him down for the count. Richard died the next day of an overdose.

I would never feel a sense of achievement when hitting someone that does not fight back. It reminds me so much of a bully beating a victim but I could not just ignore things that were eating away inside me. I would always leave someone alone once I had released my anger. Its strange how I was still able to take personal insults and challenges in my stride without too much concern until someone just pushed that little bit too far, but when it comes to my family being hurt I just go wild.

My peers in the martial arts are forever telling me that I should be at the stage in my martial arts part to be able to forgive and forget. But if an injustice is done to my family or me but I still feel fuck that, those that harm my family need to be dealt with. I cannot go round and forgive someone for harming a member of my family whilst also offering to wash their car and mow their lawn. I'm still of the frame of mind that I'm going to come charging through their street door at 5am to beat fuck out of them while they are nicely tucked up in their comfort zone. In the eyes of the law I'm wrong, in the eyes of some of my peers its wrong, but I tell you something, the personal satisfaction it gives me feels fucking great. However believe it or not, I am one of the most peaceful non-aggressive people that you could ever meet and would be a happy man if I never had a violent encounter ever again in my life.

Some situations have really caught me unaware like the time I kicked off with the guy that lived upstairs to me. He was a heavyweight African with tribal scars and a stare that would cut you in half. I don't

89

have a problem with anyone because of what they are or how they live their life until it affects me in some way, and this guy did. He was a member of an African tribal dance group, which in itself is nice to see a bit of someone else's culture. But dance practice at 3am in the morning on my ceiling was a bit much. I would go upstairs and bang on the door and the noise would stop but no one ever came to the door. This carried on for a few weeks until I finally cracked. I banged on his door but he would not come out so I asked one of my neighbours to give me a call at work when they see him downstairs by his car so I could have words with him. I didn't want to fight; I just wanted him to be reasonable about when he practised his foot stomping routines. I got the call from my neighbour and shot home as quick as I could. As I pulled up he was standing there talking to another guy who must have been about four foot nothing and looked like another member of his dance group due to the similar clothing that he was wearing. I jumped out of the car and just exploded. My plan of discussion and sensible debate went out of the window.

'What the fuck do you think you're doing fucking dancing about all night when I'm trying to sleep'

I think I was trying to psyche him out so it wouldn't come to blows because he was massive next to me and I would have had my hands full if he could fight.

Next thing I know the midget has jumped on my back like a frantic lunatic. He was biting and pulling my hair whilst heel kicking me with his legs that he had wrapped round me. If you have ever seen the Movie 'Ace Ventura-the Pet detective' where Jim Carrey fights the tiny guy that gets let loose from the holdall, it was just like that, fucking unbelievable. I couldn't shrug the little guy off my back so did a backward fall as in a judo breakfall and crushed him between the pavement, and me knocking the wind out of him. The big guy just stood there saying nothing. He didn't utter a word of make any fighting motions at all. My shirt was ripped and for a few seconds I was just standing there kind of in shock. I couldn't believe what just happened. The little guy then got up and moved closer to me whilst putting his hands up to box me. He had one leg shorter than the other and was trying to bob and weave as he threw feeble combinations of Jab and cross to my face. I let go with a palm slap to the side of his head just like you would hit a baseball, which took him off his feet and told him to stop being so fucking stupid. I couldn't fight a short-legged midget but had to deal with him in this way to stop him coming at me.

I walked off to go indoors to change my shirt and get cleaned up when the little guy shouts

'Next time I won't let you off so lightly'

I laughed and went indoors. That was the last I heard from them and the big guy was always polite to me when we ever crossed paths after that day.

It just goes to show you how you can easily misread where the real threat really is. Most people would have put money on it that my problem would have been with the big guy.

I decided to go back into music for a while and set up a new band around 1985. I was very much into stuff like U2, Big Country, and similar guitar rock bands and tried to write stuff along these lines.

I was a crap singer so took on a vocalist and was not too hot with the instruments but I wrote pretty good

Songwriting in Norfolk - September 1985

songs. I wrote about 5 songs and went into a small studio to record them. I think it cost me about £130 to record these songs in a day. It was pretty basic stuff. I had a few hundred-cassette tapes run off of the songs that we recorded and sent them out to record companies to see if I could get a deal.

A few months later I had a call from a guy who said that he had listened to my songs, was interested in meeting me and could he take me out to lunch to chat about my songs. I said that I would meet him if he covered my day's expenses and paid for the meal because I was very busy. He agreed and told me to pick the restaurant and let him know when we were to meet. I picked Mr.Chans in Soho, which at that moment in time was not within my budget, and went along to meet him.

To cut a long story short, I had this meeting with these two businessmen who offered me £60,000 to go away for a year to Norfolk and just write songs. At that precise moment I did not have a penny to my name so what do you think my reply was?

Yeah, you guessed right, I packed my bags and off I went.

The good thing about this deal was that I didn't have to sign any contracts and did not have to return any of the money, regardless of what happened. The only agreement we had was that if they could get me a major recording deal for the songs I wrote, they would get a 10% cut. They were very generous and honest with me, as I was with them.

They hired a big studio, a record producer, a musical director, and got me on the radio promoting my songs, but doors wouldn't open. The problem, as I saw it, was that outside influences and the move towards trying to get a commercial sound killed off the original raw sound that they first heard in my music. The final product was something quite different to my original sound. The year passed quite quickly and we all parted company as friends. That was it. Neither of us had any commitment or conditions that we were tied to and we all went our own ways. My guitar now gathers dust in the corner of my office; I have not recorded any songs for about 10 years now. However I recently had a call from my pal Gary Wilson from GDI Records, who was recording with Caroline Munro who you may know as the Bond girl, Naomi, in The Spy Who Loved Me and he has also released singles with actor Christopher Lee the villainous Saruman in **'Lord of The Rings'** also in **'Star Wars: The Attack of the Clones'.** He came across a tape of mine, that I gave him in 1988, liked some of my songs and hoped to record them for inclusion in a movie project on which he was working. He also wanted to record some of them with Caroline Munro, so who knows what will happen next! I'm in Gary & Christopher's DVD in America but yet to see it.

Me looking after the biggest baddie in the universe at a days filming

Back to me now in 1985.

Whilst away in Norfolk I had my family look after my flat in Barking. I would call them each week to see if everything was ok. I used to return to London once a fortnight, to take my son Jamie out for the day. This was the only access the court gave me at the time. I would drive hundreds of miles from Norfolk, collect him and take him to the zoo; McDonalds and all the usual places that Sunday fathers go, and then drive straight back to Norfolk again as soon as I dropped him off. I didn't dare miss a Sunday with him because he was my life. I didn't see the lengthy journey as any sacrifice at all.

I never had time to go to stay at my flat because I had to be back on Sunday evenings, to travel to the recording studio. I had a Triumph Spitfire at the time and the journey finally took its toll on the car. I was driving along a stretch of road, near East Rudham, about 70 miles per hour when one of my rear wheels came away from the axle. My car went out of control, over a 40-foot drop and smashed into a farmers wire fence. I don't know how I survived the crash but I did. The car was wrecked. I can still remember listening to Simple Minds on my cassette as the car took off.

When it was time for me to finally come back to London my family sent my flat keys to me in Norfolk. I thought that they were showing off because I never visited them while I was away. I soon found on my return that it was something else. My flat had been gutted. My TV, video, furniture and everything else was gone. Even the coin slot meters were broken open. They claimed that my flat had been broken into via the living room window and all the furniture etc must have been taken out that way because there was no damage to my front door whatsoever. To this day I don't know whether it was my mother, stepfather or sister that did this to me, but one of them sold everything in my flat. They thought that I had made it big in the music world and would never be returning to the flat. I know that it was one of them because they had the only set of keys to the place, plus I could not have been robbed via the window because I lived four floors up in a tower block. What did they think that I was a fucking idiot, to even consider a story like that? What could I do though, it was my own family. Two alcoholics and one junkie. I should have expected something like that. All I could do was accept the situation and start again. After I came back to London from my year away in Norfolk I had the urge to get seriously back into training. I had spent most of the year in a cottage that Dave Boy Green had rented for me and spent a lot of time with him over that year. If you are not an avid follower of boxing, Dave was famous for his fight against Sugar Ray Leonard for the World boxing championship. He had a video camera and we were always playing about and recording stuff especially when he would pop into the recording studio with bottles of wine and we would sit there having a laugh and getting pissed. He was a great laugh and we remain friends to this day. In fact he recently came all the way to London from Cambridgeshire to meet me just to be interviewed for my book 'What makes tough guys tough.' Anyway a while after my return I decided to open up a full time martial arts studio. It was now 1987. The idea was to open up a gym, which was basically style-less. As far as I could see there were full time studios running for Taekwondo, Wing Chun, Karate and so on but nobody had a studio open where martial artists, regardless of style, could come and train on the bags, weights, speedballs, mirrors etc. Without having to also take on and learn the in-house system or style of training. The reason that I wanted to do this was so that I could gain knowledge and exposure to the variety of arts out there, without me having to travel around to all the different clubs, styles and systems. I was a Karate 3rd dan at the time and had spent five of those years also training in full contact kickboxing and wanted a change. I was sick up to the teeth with karate and I was getting a little tired of taking punishment to the head from full contact kicks and punches so needed to move into another direction. I was also advised by my doctor to stop taking blows to the head due to the regularity of excruciating headaches that I was suffering from. I would literally have to lock myself away in a dark room for three days and nights before the pain would go. To this day I still have bad head pains which far exceed what you would class as a headache but I have grown to live with them. I don't have much choice really being as the hospitals and

The New Breed Academy in Walthamstow, East London 1987

specialist cannot find a cause or cure other than brain cell damage.
Anyway I set up my full time gymnasium in St.James street, Walthamstow, in East London. I called it the 'New Breed Academy.'
I chose 'New Breed' because I now wanted to create a new outlook to training where you could take anything from any fighting art that worked for you, to end up with your own tailor made hybrid system of self protection. I and those that wanted to follow this concept would be the New Breed of Self-protection instructors.
The word 'Academy' came from the Greek philosophers. They set up schools of knowledge where the Greek children would go to study and learn. I wanted my gymnasium to be a place where people could come to gain knowledge. So putting the physical and theoretical

95

aspects together gave me the 'New Breed Academy.'

I had saved up as much as I could from my job as a reporter for the East London advertiser, and put in everything else that I had and opened up for business. I made it affordable so that absolutely anybody that wanted to train and do their own thing could come down. It was open from 10am to 10pm six days a week and you could come in and train all day, every day for as long as you wanted for only £5 per week. Can you believe that? £5 per week.

I was not in it to make money. I just wanted to cover the bills and gain as much knowledge as I could. I spent around 100 hours every week for a year just training, learning and teaching. The martial arts and fighting related knowledge that I had exposure to over this period would have taken me a hundred years to find and learn, had I travelled the globe to see it all.

I met some of the biggest tossers that you could ever come across in your life but also met some of the best people that you could ever wish to meet. The Academy stirred up a lot of interest throughout the U.K. and made my name known within the martial arts community. At this time my first ever articles were featured in 'Fighters magazine' under the headings of 'Which system of training will help you most' followed by 'How to put together your own fighting system.'

This was the beginning of me becoming known for airing my views on self-protection and the need for us all to have a different outlook towards our systems of training. What I'm going to say next will sound like I'm being very flash and big headed but I cannot think of another way to put it. I was ahead of my time. The U.K. martial arts world was just not ready for me. All they could see was a 26-year-old, mouthy, karate man trying to educate them about what's real and what's bullshit. What could I possibly know? What could I tell these wise and knowledgeable people? They basically put their fingers in their ears and closed their eyes hoping that I would go away. It took a few years to pass but just as they thought it was safe to go back into the water they unplugged their ears and opened their eyes to get BOOO! My good friend Geoff Thompson was there to greet them and blow them away, and a mighty fine job he did of it as well. Geoff along with his colleague Peter Consterdine did for the martial arts scene, what the punk rock explosion did for music. It opened up the doors to a whole new world, allowing a new breed of martial artist to emerge and shine through.

When I tried to break through in 1987 I was just at the wrong level at the wrong time. Although I had the ability and knowledge. Nobody wanted to hear it from a 26-year-old. Just as nobody wants to listen to an 18 year old policeman. For something to be accepted it has to come from the right people at the right time. The chemistry between Peter and Geoff was just that, it was the right time. The climate and progression of the UK martial arts scene needed its punk rock explosion and they made a superb job of it.

Anyway, let's get back to the Academy in Walthamstow. I had reached a stage in my life where everything that I wanted from the academy had been achieved and I had to now either run it as a proper business or get out because the bills were starting to pile up.

The trouble with me is that everyone who trains with me ends up becoming good friends of mine. I was never into the role of being the high and mighty martial arts instructor, who kept his students at a distance, treating them like you would your pet dog, expecting them to respond to each instruction that you make. I treated people like I wanted to be treated myself. The problem with this is that close friends tend to not want to pay their way, hoping that the cost of running a gymnasium will be met by the others. All it takes is half a dozen people doing this and you find that no money is coming in. As a result they lose their place to train. Another thing that was happening around this time was the seriousness of my custody battle for my son Jamie. He was then 5 years old and still an obsession for me. He was my life and I did not want to miss the experiences that come part and parcel in raising a child. I wanted to give him everything that I never had and protect him from all the shit that I had been exposed to whilst growing up. I also needed to get hold of some serious money in order to pay for my solicitor, private detective, and all the other stuff that you get involved with when dealing with courts.

While all this was going on more and more of Karen's junkie friends were attaching themselves to my mother with their hard luck stories and craftily befriending her. My mother was very weak minded when

97

she had been drinking and the junkie brigade could see that and would take liberties with her. They would steal from her, use her phone, and borrow money and food, which they never returned and basically walk all over her. I would go round there and throw them out and explain to my mother that they were just vultures attaching themselves to her because she was weak, but it is impossible to talk sense to anyone that is constantly under the influence of alcohol.

In the end I had to book a holiday for her to make her go away for a couple of weeks, so that I could lock her flat up and get rid of all the hangers on. I paid for her to go to a holiday camp for two weeks and hoped that things would improve when she returned. When she came back her flat had been robbed and it had been used as a squat and drug den. Nobody was around when she returned so the police couldn't do much about it. My mum seemed to realise that it was time to stand up for herself a bit and refused to give in to sympathy stories from Karen's friends. Unfortunately though for my mother, she had let one of the junkie crowd know that she was putting a few quid away each week so that she had money at Christmas to buy gifts for all the grandchildren. One of them set up to snatch her bag when she collected her money from the post office. This we didn't know until it had all happened.

As she entered her tower block and was waiting for the lift to arrive three people with hoods covering their faces confronted her and grabbed at her bag. She put up a struggle, which was probably more to do with Dutch courage, from being drunk, rather than a conscious decision to protect herself. The muggers beat her up, knocked out all her front teeth, cut her hands legs, and head and ran off with the bag, which contained her Christmas savings.

Tears of anger and frustration came to my eyes as I received the call from the police officer at the hospital telling me that my mother was being treated after being beaten up.

It is such a hard thing to deal with, when you are capable of tearing another human being apart with your bare hands, and something like this happens to you, where a member of your family gets hurt and abused and you cannot do anything about it. I wanted to get these people and inflict the most horrible form of torture that I could possibly devise. When it comes to a member of my family getting hurt I just go crazy.

I go beyond any form of sensible reasoning and acceptance. All I want to do is make the perpetrators suffer.

I tried my hardest to find out who it was or who had set it up but nobody knew anything. Everybody knew that I meant business and that someone was going to pay dearly for this mugging but nobody was talking. I even offered to pay a few grand, in money or in drugs, to anybody that could give me a lead, but got nothing. I don't have any sort of dealings with drugs but if it came to it I would have found a way. To put it mildly, I was angry! I eventually found out who did it. Things always rise to the surface in the drug user world and when it did I was there to deal with it. There is a saying that time heals but this doesn't help my mother who was permanently disabled due to their actions until the day she died, also, unfortunately for them, I am not a man of forgiveness. They will never hurt another person again.

So with all this combined with all my own problems, the New Breed gym had to go and I set up a new business dealing in corporate entertainment. A new craze had just come over to England from America called Paintball games. It consisted of two teams of players dressing up in camouflage clothing and safety goggles, then being let loose in the woods to shoot each other with plastic guns, powered by CO_2 capsules, that fired paint balls which exploded on impact. It's not something that I was into but the business opportunity was phenomenal. I was taking around £1,000 a day. I only worked two days a week. If it could have been run everyday I would have done it but it was something that appealed to the yuppie market and they all worked Monday to Friday so would only book a day at the weekends. In the weekdays I would do different types of security related work because I knew that the paintball boom would only have a year or so before it died out so I had to keep my self involved in other areas. The other good thing about the paintball business was the fact that I had all week available to look after my son if I gained custody of him. It took two years of going back and forth to court, with nothing but disappointment, until I finally won the case. On the final week when I was due to go to court I had bought a telephone recording device from Tandy's for about a £1.00 and recorded every phone call I received. The reason for this was that unlike the normal anonymous threats that I had always had in the past, these people clearly identified themselves and who they were working for. I was told quite clearly that if I went to court for the final decision, whether I won the case or not, I was going to lose my legs and a member of my family would be opening their street door one day to receive a face-wash of acid. Can you imagine how that felt! My stress levels went way off the chart, I was becoming very ill, weak and run down with the worry of it all. What made it even worse was the fact that I had a gun with 650 rounds of ammo and was so hyped up that I was ready to blow away anyone that even remotely looked to be a threat to me. I had really lost the plot and was not an ideal person to be in possession of a gun at that time but I fulfilled the home office requirements and was legally licensed to possess firearms.

It had even crossed my mind on many occasions that the encounter that I had in Barking when the 3 guys dragged me into the alley could have been connected to this recent threat. I was going crazy and I was scared. The tapes of the telephone threats were given to my solicitor to hold in the event of anything happening to me.

On the day of the court decision I made sure that the members of my family were safe and sound. My stepfather Peter wanted to go to court to support me but I told him I wanted to be alone to deal with this. Peter knew that he was dying and wanted to shoot the people that were giving me all these threats. The problem was that he would have done it as well. He was in the frame of mind that, if I were the only living relative of Jamie's alive, I should automatically get custody. At this time, Jamie was being looked after by his uncle, who had cunningly taken on the role of his guardian. It was a very ugly and heavy situation all round with some very serious threats being made. I was the only one not making any threats, putting silly things into writing or having myself being recorded. I played it really cool and poker faced. Inside though I was a nervous wreck. My right hand man in the martial arts world, Bob Wallace, came to court with me. Bob knew everything that was going on in my life and was someone who I trusted. Bob was your typical hard arse Northerner who came to London and came across me. He had been in the forces, seen a bit of action and was finding it hard to fit in with normal civvy life. He had done some martial arts training, wanted to get back into it and came to train with me. He became a good friend and regular training partner. He was also a fucking nutter. Bob didn't give a shit about any threats or peoples reputations. He came to court with me to give me a bit of support and also watch my back but it was just a shame he dyed his hair bright white the night before and arrived drunk. It just didn't look too impressive being as he

Peter at the early stage of his illness

was supposed to speak up on my behalf. At court, the stupid thing is, everyone sits in the same area waiting to enter the courtroom so it was very easy for people to intimidate and psyche each other out. I was hoping that Bob didn't lose his cool through being pissed and kick off because that was exactly what they were trying to make us do. They taunted us, spat at us, and used threatening body language and gestures to try and intimidate us in any way they could. I knew and they knew that I had a good strong case, which could tip the scales in my favour this time and they were worried Within an hour I was walking out of the court and had been granted care and control of my son, which meant that he was now coming to live with me. I had done it! A father winning a custody case was a rare thing. Things became heavy big time after that so I basically disappeared off the face of the earth. I moved all over the country so that I could not be found. I just wanted to get on with my life and was sick of all the threats, phone calls, wreaths, car damage, hate mail etc, so off I went. I broke links with all my friends and just moved around for the next four years until things finally calmed down. I was not able to do normal 9-5 jobs because I had to get Jamie to school and be there for him when school finished, so I had to make money in anyway I could. Many people from the martial arts seemed to be doing some form of security related work. I don't know why this was because nothing within martial arts trains you for security work, but, for some strange reason many of us drift into this field. It must be the fighting connection. The same goes for people from the forces. They also seem to be drawn to security related work. I would think that they are probably more geared up to do security stuff due to their training but having never been in the forces myself, I don't really know. What I do know is that the martial arts don't teach you shit about security work. My security side came as a natural progression from my mod days looking after bands and stuff like that. My stepfather was nearing the end of his life. He was now housebound and had to sit with a bottle of oxygen beside him at all times just so he could manage to get through the day. My mother, Karen and myself would all visit him at different times so that he wasn't on his own. His biggest fear was dying in his flat alone and he knew he was dying.

He still carried on drinking scotch and smoking his Old Holborn rollups, which he knew was shortening his life, but it was the only life he had. The man that I once hated so badly who hurt me both physically and mentally was now hurting me in a different way. He was weak and frail and I had grown to care for him. He was my mentor and only father that I really knew, I could see him gradually being taken from this world before my very eyes. The thing that made it even harder for me, is that I have no religion, so had nobody to pray to and beg for more time with his life. There was nothing. This was very hard for me. I had nobody to turn to apart from myself. I was hurting badly. I'd spend all my available time just sitting and listening to all the things that he had been up to in his life and found areas of his life that I had never known about before. I asked him if I could write down his stories, so that we had a part of him to pass on, to mine and Karen's children who would surely forget him by the time they grew up. He agreed to help me record all these facts and we began writing down his story. I bought a Dictaphone and loads of tapes and let him just sit there each evening and talk away. He would talk about his younger days as a 16-year-old, when he broke into the Chinese Embassy at Regents Park, which he described as a grand house, out of this world. He said that he didn't know it was the Chinese Embassy, it was late at night, dark, and to him it was just another big old house, but once inside he knew that he was somewhere that he should not be. There was nothing to steal. In no time at all there were police everywhere.

He was sent to the Magistrates court and ended up at the London Sessions, receiving 18 months in Wormwood Scrubs.

He also loved to talk about the days when he began rubbing shoulders with Reggie and Ronnie Kray, they had a club called the Double R at Bow road at the time which was their first ever club. He said that they were

Peter (Left) 1958 in East London

quite nice guys who didn't really do any villainy themselves, but were at the strong-arm game. You know, protection and all that, but nothing really heavy, then they got the Kentuky club in Mile End road, so with drinking in the clubs and working in those circles he got to see quite a lot of the Twins. Also around that time he was working with a bloke called Buller Ward who he said was not really a villain, but again was well into the protection game. There were around 60-70 of them, all in and around Bethnal Green, Shoreditch, Hoxton, Hackney, etc .

I found it so fascinating, listening to all his adventures, but I don't see how any of them were not villains. I think that he was just playing 'Old school' and protecting the names of his own kind. It gave him a real feeling of enjoyment to be doing something, rather than just sitting in his chair waiting for his day to come. The stories piled up and it was easy to see that a good book was in the making. I started to structure the material into book format and had a brilliant book on East End villainy, which included his part involvement with the

103

Krays, Buller Ward, Golly Sillet, and many more known faces that he had worked with. He was beginning to deteriorate rapidly and had to be taken into hospital for round the clock care. He could not be cured but his life could be extended by a few weeks.

My son Jamie was now in Senior school and understood what was going on. He asked if he could come to hospital to see his granddad on my next visit.

It's a hard stage in your life when someone close to your children is dying. You have to decide what to tell your children and decide who should be shielded from the pain and stress of worry that comes with illness and death. I felt that Jamie was old enough, so I took him, but I didn't feel that my other children were ready for it.

That was the last time I saw my stepfather alive. My sister called me from the hospital; to tell me that the hospital contacted her, to say that Peter was in a bad way, by the time she arrived Peter was dead. He died alone.

I have never felt such an emotional pain like this ever in my life. If you have ever lost anybody close to, you will know what I mean. I did not know it was possible to hurt so much. The strangest thing about it all was that all the suffering that Peter had put myself and my mother through over the last twenty or more years did not affect my feelings at all. I so badly wanted Peter to be alive again. Shortly afterwards I went to the chapel of rest at the hospital to see Peters body, once more, before he was cremated. I don't know if I went for my own reasons, or if it was to give my mother support, but I went. I just broke down and wept. Any toughness, or big man attitude that I had, was just drawn from me. I was reduced to a weak sobbing child. I was in my early thirties but felt like a 5-year-old, unable to control my emotions. It was a very hard time in my life. From that day on I was never able to look at a photograph or anything that reminded me of my stepfather. I had to block it out of my mind to stop the pain. My mother and sister dealt with it differently, as we all do when dealing with the loss of someone, but for me I had to blot it out. Time passed and my younger children were still asking when they could go and visit their granddad again so I had to tell them that he had died.

It took five years to pass before I could look at the photographs that I have included in this book.

Back into Training

Martial arts are strange. No matter what other problems you may have in life, no matter what other interests you have, you usually always creep back into the arts in some shape or form.

I was kind of like the Sunday footballer. I would get together with the lads and have a kick around once a week and then leave it alone and get on with other matters in my life. The obsession that I had with the arts in my youth had gone. I was content to just play around with the arts as and when I felt like it. This is the same for most other people that have been in the arts for a long time.

Me (middle row 2nd right) – Black belts - cross training 1982

You simply get lazy and bored. Martial arts had lost its buzz for me. My main problem was that the further and deeper that I searched for reality within the arts, more and more of it became bullshit to me. Most of the instructors out there preaching self-defence didn't have a fucking clue. They had never been in a real street battle but were preaching combat philosophies and selling their psychomotor movements to others that didn't know any better. These students were, in turn, becoming instructors themselves and were the next generation of bullshitting street fighting experts. It was a real joke. There were some good guys out there who would shine through but most of these guys were very low profile. Martial arts were getting a bad name because there were so many pretenders. This has never happened in boxing. The guys that get in that ring do it for real.

They punch fuck out of each other until one of them gets a result. Yet you never saw boxers opening up self-defence clubs. All you would see were guys in white suits wearing black belts hoping that they would never get in a real fight.

I became so sick of the martial arts that I stopped mixing with them. I was still teaching what I considered to be the real stuff, but then again, everybody thinks that his or her stuff is real!

After my mother got mugged and beaten it made me think more about self-defence. Who was teaching people like my mother to defend themselves against the scum of society? The supposed experts were ignoring her and others like her, the weak and the vulnerable. Most martial arts instructors were only catering for the 20-year-old fitness fanatic, who would be able to put up a fair defence without even knowing the martial arts. So we had black belt martial arts instructors, claiming to be street capable, teaching young adults that were in their prime of health and fitness to defend themselves! It was all arse upwards. I changed my outlook as to what and whom I was teaching it to, ending up with my own method of training which I put out under the label of 'The New Breed system.' Prior to this my contact with a guy called Geoff Britton also had a big affect on what I was doing, I will talk about Geoff later. So I had a system of training, concepts and philosophies that I felt were relevant in order to cater for the type of attacker that we are likely to meet day to day. We no longer needed to be able to punch through the wooden armour of a Samurai warrior as taught by many martial arts. We were more likely to be greeted by a steel toecap boot or broken bottle.

I have had more street fights than most people have had haircuts and in all this time I have never been attacked by anybody that resembles someone fighting from horseback. I have also never been attacked by the variety of obscure martial arts weapons that many people practice their defences to and I have never been able to make an X block work against anything, other than vampires. I had wasted so many of my years training in nonsensical garbage that I was annoyed with the martial arts. I was pissed off that I had been led up the garden path for so many years with nothing to show for it at the other end. Ok I did have some pretty high level qualifications

Me (centre), Geoff Britton (2nd right) & some New Breed lads 1987

within the martial arts world but qualifications don't win fights. As a result I became very anti-martial arts at that time. I was angry that this bullshit was being allowed to happen.

At one period in my life I had 10 karate clubs running and was making a nice living out of it but I was selling misinformation. I was selling a lie. I knew that a lot of the stuff I was teaching was bollocks, but that's the way things were done then. My full contact karate coach Geoff Britton would say to me,

'Jamie the stuff your teaching looks lovely and is making you a lot of money, but do you realise its shit? You need to make contact with your blows and also take contact from blows to feel reality'

I couldn't believe what he was saying to me! I was a 1st Dan black belt and he was I think a 4th or 5th Dan at the time. I expected him to be well impressed with my students. I asked him why he thought my clubs were shit and he said,

'Jamie, you are a good instructor and a good at what you do on the streets but you are not passing that street knowledge on to your students, you are passing on shit', and he was right.

I was teaching the martial arts as it was taught to me and was blindly passing it down the line without question. I knew that a lot of the stuff I was teaching was bollocks but I wasn't doing anything about it. I was still passing on martial arts history and simulated fighting the way it was done centuries ago. Geoff was the bollocks. He set me on a different path of learning that changed my whole outlook on how I taught. I took Geoff on as my mentor and never looked back. Geoff had one of the finest stables of full contact fighters in the country and was very good at what he taught. I never had the ability to become a champion like all his other guys because I was not interested in sport martial arts and I was too much into the reality of street fighting, plus being honest, I wasn't any good at it anyway. He had about six British champions in his club at the time and I certainly was never of their standard. I think that another reason that I got on so well with Geoff was that we both understood music and had both been in bands. Geoff was much more successful though because he played drums for Paul McCartneys group, Wings.

Geoff used to teach me much more than martial arts and how to fight, he was very educated in life's hidden skills and would do things that you didn't get from your average martial arts instructor. For example he once said to me,

'I'm training Paul up for his next fight and want to do something a bit different, can you be available for training for all of the next week' I said,

'No problem, I will pop over to see you'

When I got there he had booked us on a survival course with Mick Lambert a Goju ryu man who I think may have been ex forces.

It was the kind of stuff where you catch your own fish and eat it, shit and bury it. We had a great time. Best diet I've ever been on.

You had to also be prepared to be jumped whilst you were unaware and got to feel what its like when you are taken to the ground, with a nose and mouth cover up, just like the military would take the enemy out silently. This was the environment where a black belt didn't mean shit. These type of techniques would be held on you, until you had used up your last reserve of oxygenated blood in your system and was about to pass out, then they would let go. The fancy

Survival course with Mick Lambert July 1984

karate stances and bullshit techniques were wiped out with ease. I took this stuff on board and made it an important part of my own system. Geoff taught me one of the most important things of my life. That you have to adapt your methods of training, according to your environment. A head butt might be the best thing to train in, and use, when confronted by an angry skinhead at a football match, but is not much use to you if taken out by a rear ambush attack that shuts off your oxygen and blood supply within seconds. Geoff was and still is a very wise man.

After first meeting Geoff, I stayed with him for about 10 years until he emigrated to live in Spain.

He often came over to England to see me and would sometimes creep up on me whilst I was working on the door of nightclubs just to observe and see if I was switched on.

From around 1987 I became pretty rebellious against traditional martial arts. I had spent a lot of time over the years visiting and training with different instructors from different arts plus had exposure to a lot of stuff from my Academy. I knew that there was so much other good stuff out there, which we weren't really getting exposure to, so I had to keep on learning.

The problem that I and most other people have, when they become disenchanted with only learning one art and start experimenting with other stuff, is that you become a bit confused as to which governing bodies and organisation you need to belong to. I had to keep links with the karate world just for the grading structure that they had in operation, because that was the only thing that I had Dan grades in. In reality it doesn't matter which grade you are, or with whom, but you go and try opening up a class in a sports centres or similar and you get asked the same old shit,

'Are you a black belt?'

It's not so bad now because people understand more about the arts and know that you don't have the same ranking system in all arts. In the early days though there was not so much awareness about different systems of training. So the problem existed that people like myself had to keep links with karate organisations, in order to hire sports halls and also get further up the grade ladder ourselves.

To the karate organisations, sports centres and all official bodies were still karate guys but hidden beneath we were not teaching or promoting karate at all. We were teaching biting, gouging, grappling, head butting, and other street survival stuff. This went on for some years unnoticed until we were put into a position where another local karate club from the Barking area were trying to make a name for themselves. They were rude, loud, aggressive, cocky and pissing a lot of people off within their own circles and also others outside of that. Ticky Donavon's Ishinryu karate were the big wigs in the karate scene in Dagenham at the time and this rebellious club seemed to be going all out to piss off Ticky and all his lads. I think some of them were former Ishinryu instructors and students who had parted company with Ticky over politics within the fighting arts. They wanted to make a name for themselves and seemed to want to do it by pissing everyone else off. For some reason they decided that they were going to challenge my club. They had got it into their heads that we did not compete in traditional karate tournaments because we were shit and didn't want to get shown up. Nothing could be further from the truth. We were training in full contact kickboxing and also studied all the other ugly violent street related stuff. To outsiders we still appeared to be traditional karate guys, marching up and down playing with the martial arts, but hidden behind that we were a bunch of real, aggressive, hard, bastards just doing our own thing.

I really couldn't be bothered with all this shit until one day a student from this cocky karate club came down and openly challenged us. These days I would laugh it off and tell them to go and grow up, or would nut them. It all depends on how I feel on the day, but at that time, I was around 26, and very much into defending my pride and took on the challenge. We didn't really have any choice then. It was that or my guys would end up taking it upon themselves to defend our name and would kick off with them in the street.

The date and venue was set and we had two weeks in which to prepare. It was agreed that it would be a semi contact tournament, with no foul play. These were karate guys who would put on a pair of boxing gloves but still do normal traditional karate, thinking that they were full contact fighters. We were very competent full contact fighters who were fucking good scrappers as well, so the semi contact rule was to their benefit in my opinion. I just wanted to get the whole stupid thing over and done with; this was not my sort of thing but my pride forced me to do it.

It was set up so that ten of their guys would fight ten of our guys and at the end I would fight their instructor.

When we arrived at the venue they were already there waiting for us. They had on matching tracksuits with white karate suits on underneath with their dangling black belts. As they saw us walk in they started a sad attempt at shadow boxing. If you see a pure karate man attempting to shadow box, like a western boxer, you will piss yourself. They were so stiff with no flow or rhythm. They began staring at us like we were their fucking worst enemies. They were snarling and making aggressive gestures. I had my old pals Bob Wallace, Bob Peters and Pete Chamberlain with me who were going to be part of the ten fighting, I said,

'I think that they are going to try it on today; they look like they have come here to kill. Just take it easy and stick to the rules, don't play their stupid game'

Bob P was the first man out and faced his opponent. Bob stood at over 6 foot and he was a big lump. The guy facing him was bouncing up and down like he had springs in his heels and as soon as the whistle blew he launched a full blast reverse punch into Bob's solar plexus. Bob hit the floor straight away. He was fighting to get his breath back and had to come off the fight area. The other side were cheering and laughing at us. I said to their instructor,

'What the fuck was that all about? We agreed on semi contact and your man blasted out a full on blow, what the fucks going on'

He came back with,

'My guys are trained for full contact, its very hard for them to pull punches, why don't we just do away with the rules and go all out to see who is the best' I replied,

'Don't be so fucking stupid, people will get hurt. Just get your guys to play by the rules or we are leaving'

They obviously took this as a sign of weakness. They agreed to stick to the rules but next

A slimmer me 1984

match their man punched one of my team members, Dave, full blast

112

in the mouth pushing his teeth through his lip. That was it, it kicked right off and Dave lifted the guy off the floor with a groin kick and then beat fuck out of him. An ambulance had to be called to attend to the guy. The next fight went on after some delay and their man tried to beat fuck out of Bob Wallace. Bob doesn't take any shit and lifted the guy up from the floor with a karate style sidekick knocking him into the trophy tables smashing the trophies and table. After the fourth fight we had to stop. We had three separate ambulances on the scene, which took three of their guys away. It's not the way we wanted it to go but they put our backs up against the wall, with no means of escape, so we had to fight. For years after that we remained their enemies, but they never fucked with us again. My pride felt good.

If only they would have stuck to the rules, they could have walked away victorious. Semi contact was not our game, but they played dirty and dirty fighting is something we were pretty good at. You should never try and beat someone at their own game if you are not proficient at it. You should find another way to do them. There's a way to do everybody.

Bouncers and Security

I have always been involved in all forms of security, from debt collecting to static watch. I did anything and everything to make ends meet. With the debt collecting though, I was not into the type of work where you would go round to a house to collect a debt from a family that's having it hard. That sort of work is for sad bastards that enjoy taking the last penny from people that have nothing. That kind of debt collector goes hand in hand with the thuggish, door-to-door, loan sharks that encourage housewives to take out loans to pay other debts. Knowing full well that they are going to struggle to pay it back, so they inflict a bit of fear to get their money. These are the shit of the earth debt collectors. I think the only thing worse than that are the Bailiff brigade. I was more into collecting company debts from business people, who were knocking people for what they owed them and getting away with it and collecting personal debts where a situation had occurred between two parties and one of them paid me to resolve the situation.

As well as this work there were offshoots of work, that were security related, but if you analysed it we were not much better that the thugs we were dealing with.

One of the main areas of work was working for a legitimate security company who supplied bouncers to clubs. Our role was to act as a mobile unit driving around London and basically being on call, if a situation occurred in a night-club, that was a bit to heavy to be dealt with by their in house bouncers, or for some reason the club bouncers did not want to be involved with it.

Situations with drug dealers and their gangs was one of the reasons that the club bouncers did not want to get involved themselves, because of the comebacks. So what would happen is, we would get a call saying that a gang of drug dealers were operating in a certain club and the manager wanted the situation resolved. We would enter the club discretely and observe. Once we found out who the main man was we would wait for an ideal opportunity to drag him into the toilet, backstage, fire exit or whatever was most appropriate and basically beat the fuck out of him, and his workers, then leave them in the alley for the police to arrive. A cable tie around the thumbs and feet

114

ensured that they did not run away and also made sure that they were found still in possession of all their drugs. I have kind of simplified the operation just to give you an idea as to how we operated but things did sometimes become much heavier that that. This system of operating kept everyone happy. The clubs were cleaned of drug dealers, the managers had a nicer atmosphere in the club and the bouncers did not have to get involved in the heavy side of dealing with drug dealers. Even the police were happy to find a drug dealer gift-wrapped with cable ties still in possession of all his drugs.

We were in and out the clubs like lightning practically leaving no trace of our visit and no witnesses. This sent a message out to the drug dealers that they should find somewhere else to operate.

Most nights we would get a basic £100 each just for driving around and being on call. If we had to deal with a situation then we would benefit by receiving a bonus. It was a good set-up, which lasted for about 5 years with one or two new faces, but generally it was always the same four or five of us. The reason it worked so well was the fact that none of us were into drugs for our own individual reasons. It didn't mean however that we were all angels either. We all had our faults and were obviously breaking the law by beating drug dealers up and restraining them with cable ties, but we didn't hear anyone complaining about it.

I had a call from my boss saying that a drug problem was happening in a club in the West end, where his guys were doing the security. The problem being that he suspected one of his own men, 'Big G', as being the distributor of the drugs. 'Big G' was head doorman at the club who was getting a fair wage to work the club and was also making a nice bit from fiddling the gate money. This happens everywhere so most managers learnt to accept it as long as they were well in profit and security were doing a good job dealing with any problems without attracting the police. 'G' managed to control in-house problems quite well. I say this because it was one of the clubs that we hardly ever got called out to, where violence had escalated beyond the capabilities of the club bouncers. It was in 'Gs' own interest to not have the police call on the club because, as my boss suspected, 'G' was running a drug operation in-house. Although most of the bouncers within the company knew of the mobile unit and how it operated, not many of them knew us that well as close friends. That's the way we liked it, because we never knew when we were going to have to sort out one of our own people that had turned bad. The way we saw it was that the bouncers, security, door supervisors or whatever you want to call them, are put into a position of trust just as the police and other professionals are. If they abuse that position, by becoming part of the scum they are supposed to be opposed to, they in turn need to be dealt with.

Anyway we decided to pay a visit to 'G's club to check out what was going on. After a while of watching his little operation we sussed that it wasn't actually him that was doing the dealing. It was someone else. But he was obviously turning a blind eye to the operation in turn for a drink. We approached 'G' and introduced ourselves but did not let him know that he was under suspicion. We said that we had received a call giving the description of four people that were carrying and dealing in drugs. Because it was his club and he was head doorman, we were going to let him drag the offenders out the back doors where we were going to give them a beating. That put him well on the spot. He absolutely shit it. What could he do? If he dragged out the drug dealers it would all come to light that he was partly involved so we were really interested to see how he was going to worm his way out of that one.

'I'm the fucking head doorman here; no one tells me how to run my club. Anyone that tries to go through these doors has to go through me' He said.

'K', being one of the lesser stable members of the mobile unit, grabbed 'G' by the hair and pushed his face through the glass door of the club, before cracking his head open on the outer wall. 'G' was lucky to have not lost his sight that night.

He works for another company now telling people how his face ended up looking like a road map when he kicked off with a mob of blokes who were knife fighters.

The way that 'K' dealt with 'G' was a bit extreme to say the least but he just had no self-control when it came down to someone challenging him. He was frightening to work with sometimes because you never knew when he would just go that little bit too far. He would never give anyone the benefit of the doubt. If there were just

116

the slightest hint of someone giving it the large one to him, that would be it.

I was working with 'K' one day when we were called to pay a visit to a rave in a disused warehouse. Someone was selling a powder substance to the youngsters under the guise of Speed. The guy selling the drugs was doing it quite openly as if he didn't have a care in the world and acted like he didn't give a fuck about anybody. It turned out that the bouncers were a bit intimidated by him because he was a bit of a name in that area, so they were a bit worried about the possible repercussions of sorting him out, this resulted in us being called out. I know some people think that if bouncers are intimidated by some people then they are in the wrong job. But to be honest, who in their right mind is going to deal with a heavy drug outfit and then be stuck on the same club door night after night just waiting for a revenge attack to come, all for £50 a night?

It wasn't a problem for us because we were in and out, never to be seen again, we enjoyed our work and were well paid for it. Where else would someone pay this sort of money to a 30 year old?

Our plan of action with this outfit was to pretend that we wanted to take over the club drug operation ourselves, so that the dealers would see us as their own type of people rather than security working for the club. Reason being that the bouncers were very concerned about repercussions but knew that this would not be an issue if it appeared as if two rival gangs were going to war to take control of the club.

Within an hour of being in the club the rivalry started and the death stares began. We made a call to 'H' who was part of the mobile team responsible for carrying the tools of our trade. We rarely used them but sometimes it became a necessity. For legal reasons I can't enlarge on that, but I'm sure you get the drift. We had planned in coaxing the dealers out of the club into the rear car park, where we were going to give them the option of moving on to another club, or kicking off with us, but things didn't work out that way. There were three of us in the club and another two on their way who would meet us outside with the tools. There were five of them, which we were aware of, but they didn't look like they were going to create us any problems. Next thing I know 'K' had caught a glimpse of this big, fat, guy from the gang giving him the big wanker sign. That was it, 'K' walked over there and nutted him repeatedly. Fatty just collapsed on the floor and 'K' stomped on his hands, breaking every bone that he had below his wrist. With that the other gang members made their way to 'K' picking up bottles on their way, while 'B' and me met them for a head on clash. We fought like wild animals. When you realise that your life is in real danger and could end within the next few seconds you do all you can to survive, you turn into a fucking rabid dog. You are unstoppable.

It all seemed like one of those American wrestling cage matches, where everyone gets in the ring and all pile on top of each other, but ours was no game. 'B' got rammed in the side of the face with a bottle, which opened him up badly. 'K' lost two teeth from a stamp to the face, whilst I took a bar to the head and got stabbed in the gut, but they ended up far worse off than we did. They just lay in a heap on the floor looking like they were the remains of a slaughterhouse. The blood was splashed everywhere, in fact even when you tried to stand up you would slip over on the blood soaked floor. The fight was over. Shortly after 'H' and 'D' arrived with the tools and we gave the drug gang the option of being introduced to the tools of our trade, or disappearing out of our faces forever. They went, but not before handing over the £3,000 in cash that they were holding, which we took for medical expenses. The drugs they were selling turned out to be Ajax and baking powder mixed together. We all took a fortnight off work and just chilled out, got our bodies patched up and had some fun.

I had a call one night from Jimmy who used to be part of our team but pulled out because he wanted to get into something a little less violent. He became a driver and minder. He would drive business people about, which was generally risk free. Jimmy called me because his main client was thinking of getting rid of him, because no problems had occurred in the last year and he thought it was not cost effective to keep Jimmy on. Jimmy was getting on a bit and definitely too old to be getting back into security work, he would find it a struggle to survive if he lost his job. He told the ladies that he was 28, but he was 40 something.

Jimmy asked if I could get him some work somewhere if he got sacked, all our stuff was a bit too heavy for him to be dealing with, but I had a better idea. It was obvious that Jimmy liked his current placement as driver/minder, so all we had to do was find a way for him to keep his job! I set it up that I would be drinking in a pub where Jimmy and his boss were going to be. I would pick a fight with Jimmy's boss and in turn Jimmy would be the hero and give me a few digs and throw me out the pub. All quite harmless and Jimmy's position would be justified so he would be kept on. Any type of security scare wakes people up to the fact that they never know when they are going to need cover. I took 'B' with me because his face was pretty well scarred up from the bottle he took to the face, when it kicked off at that club, so he looked really mean and hard. It was all to make it look better for Jimmy. We were sitting in the pub when Jimmy arrived with his boss. We let them settle in and then it was time for action. I walked past Jimmy's boss and barged him to make space to get past, saying,

'Get out of my fucking way, I'm trying to get to the toilet, do you think you own the fucking pub?'

With that Jimmy grabbed me and pulled me away from his boss and in a snarling voice said,

'If you've got a problem, make it with me', then he nutted me.

It was pre-arranged for Jimmy to stick in an uppercut, which I had tensed up and prepared for, but a head butt I was not ready for. Although my nose was as flat as a pancake from being broken many times before, Jimmy's butt still opened it nicely and the blood was running all over my face. 'B' was over at the table pissing himself with laughter, because, instead of Jimmy working to plan he just reacted naturally, as he would if it was a real fight, and stuck the nut in. What a stupid wanker! Then, as I put on the angry psychological act and was screaming,

'Come on then, outside if you want some!'

His fucking boss bottled me from behind then ran. Jimmy pushed me to the side and went out after him. His boss had jumped in the car and locked the doors. He let Jimmy in and they drove off.

I went back over to 'B' who had tears coming out of his eyes with the laughter. He said,

'Is that Jimmy a fucking wanker or what? if he had half a brain he would be dangerous', he just couldn't stop laughing.

119

I was wiping my face and rubbing my head, where he smacked me with the bottle. Luckily it just bounced off or I would be having stitches put in, however, it did leave a nice bump.

I just sat down with 'B' who was still laughing, shook my head and said, *'He's a fucking wanker'*

Jimmy phoned me later that evening, apologising for fucking up, and arranged to take me and 'B' out for a meal and piss up. We went out and had a good laugh about it. We went through the whole event step by step, repeatedly, until we were too pissed to speak any more. Jimmy's job was secure again and I had helped a mate out in a time of need. He said I could call on him if I ever needed a favour, which I knew anyway. That always went without saying with our lot. We were very loyal to each other. My time working with this outfit was financially one of the most rewarding times of my life, at the expense of other areas of my life. I had fought in court for seven years to gain custody of my first son, which I eventually won and raised him from the age of seven to sixteen. Throughout this time I had been through another disastrous marriage, which brought about another three beautiful children, who, in turn, were taken away from me. So I was now able to see my first son day and night, but my other three were reduced to seeing me twice a week. I just couldn't get the balance right. I was now raising my son on my own and knew that I had to have a change of career which wasn't so life threatening and allowed me to care for my son. I drifted gradually back into doorwork, the nearest I could get to a regular job, which was just like working nights in a factory. My mother looked after my son in the twilight hours and I was always there to get him up for school and cook his meals in the evening. It became even easier as he grew older.

Working as a bouncer was a job that I hated but it was regular money and stopped me becoming a hermit. The licensing of door supervisors was now being introduced and was causing many bouncers to panic, because of having criminal records, so they were holding onto the positions that they had, hoping that they would slip through the net or stay just outside of the jurisdiction of the councils. I too avoided the scheme for as long as I could but for different reasons. I knew that I would check out ok as far as convictions went, because I was a licensed firearms holder, but didn't like the idea that your badge had to be displayed for the public to see. The problem that I had with this was that the licence was openly displayed on your chest and it had your individual registration number on it. That in itself doesn't sound too drastic but the problem here is that you can take any doorman's licence number, go to the local council where the licence was issued and you will find the doorman's home address, available for public inspection. So in theory every customer that

The dreaded door supervisor's badge

you have a problem with, at a venue, can if they wish pay you a home visit or even worse, pay your family a visit while your out working the doors. Can you image a police officer, or military personnel having to wear a badge that could indirectly give away their home address. In the eyes of the council, they liked the licence number being visible on the door supervisor because it supposedly would make it easier to identify the bad guys. If you received a kicking off a bouncer for no reason at all, you could take down their number without them even realising it and have them reported to the council or even arrested in no time at all. Another problem with the badges were that if it did kick off and it ended up in a rugby scrum, somebody would rip the badge off your chest and then have a nice passport size photo of you along with you home details. If a situation ever went to court, witnesses that had never seen you in their life could still pick you out, just from remembering your likeness on the photo. This might seem a bit extreme to you but believe me, it goes on. So I avoided it as long as I could but eventually had to get licensed. We now had a system set-up where you were seen as a proper doorman if you had a badge but a cowboy if you didn't. This of course was not true because the licence thing was a load of bollocks. All it did was rid convicted villains from the front line of visible security, which made it look all lovely, organised, and professional, but this also created a new monster.

The older, experienced doormen were slowly being replaced by young, inexperienced, 'Jack the lad' doormen who would work for half the price of what the older guys would. So the pubs and clubs were getting twice the amount of security for their money, but getting nothing more than a bunch of inexperienced bovver boys. Along with that came young doormen who would also be selling and taking drugs. They were on shit money, so saw the drug scene as a good earner, plus they would be able to confiscate and use any drugs that they found on other people. It was a gold mine for young security guys, which, in a crazy sort of way, was helped by the licensing scheme.

Working on the door means different things to different people; depending on what angle you look at it from.

From the general publics point of view, the bouncer, or doorman, is a guy, who is supposed to stop trouble, but, in reality a lot of doormen abuse this position and some do create the trouble themselves. I have worked with some really good guys, who do a good job of security, but I have worked with ten times the amount of tossers, who are as corrupt as fuck.

If there is trouble in a bar or club, people tend to look to the bouncer to stop it and sort the problem out, same way that they would call a policeman in the street if they had a problem, but what happens when you call the bouncer and he is the main ringleader of the drugs being sold? . I saw this happen in the 'Cable club' when Freddy was called to deal with a complaint about someone offering drugs to a customer. He was always a bit paranoid, from doing speed himself, and suspected everyone of being undercover police. He grabbed hold of this guy that made the complaint, and head butted him about 5 times. The guy was a mess. I was in the club as an observer, having a drink with a pal, when it went on. The police were called and Freddy spent the night down the nick giving a statement. Due to the fact that he was hyped up from, the drugs he had taken, he gave the police more than enough evidence to put himself in it. He got two years in prison and the story goes that he made a nice wife for someone in there.

The Management of clubs rely on the bouncers to throw out the people who are causing problems in the clubs for the normal things, like creating violence, stealing handbags, selling or doing drugs and those who are excessively drunk. But not for informing security that someone is selling drugs.

Drunks are the worst problem though. It is an offence for a landlord to sell alcohol to someone who is drunk, or to have a drunk on the premises. Even though it is the same landlord who sells you the drink in the first place. Once you are drunk, the proprietor can be prosecuted for it, so they do not really want drunks on their premises.

It's then the job of the bouncer to get the drunk out of the club before they start making a nuisance of themselves. The problem gets even worse when a drunk collapses in the cubicle of the toilet with the door locked. Apart from having a dump, the toilet cubicle gets used for taking drugs and maybe the quick bit of passion. We were always chucking couples out of the cubicles, who were pissed and having a quick shag totally switched off to their surroundings. When I worked in Soho we found all sorts of stuff going on in the toilets that you just wouldn't believe. But as I said the drunks are the worse. One time we had a call from the manager, saying that somebody had locked themselves in the cubicle and he thought that they had overdosed on drugs. I went with the other guy, 'O', that I was working with that night, to deal with the problem. As you know, men's toilets are a real shit hole, they always stink of piss and you have to tread careful because the floors are always covered in piss, phlegm and sick. If you got any of it on you it was there for the rest of the night. You could wear rubber gloves and wash your hands but you couldn't wash your clothes.

I looked over the top of the cubicle and saw this guy laying in a heap on the floor, covered in sick. He had spewed up on himself and was talking in a drunken state. I reached down and unbolted the door but we still could not get the door open because he was propped up against it. I then had to reach down and try pull this guy upward and backward so that 'O' could squeeze in the gap of the door. At times like this it is quite handy being a bit of a 'lard-arse' because there is no way that I could have squeezed in the gap, so 'O' had to do it. His job was to push the guy back and prop him up on the toilet seat so that the two of us could then help him out the cubicle and send him on his way.

As he grabbed the guy at arms length to prop him back I let go, they both slipped, with the sudden increase of weight, and both went over. The drunk managed to produce the last helping of his night's meal and beer over 'O' who was well pissed off. He even got some in his mouth. 'O' reacted with a right cross to the drunks nose so we had blood all over him as well now. 'O' walked off in a rage.

With the help of another guy, we eventually got the drunk out of the club and called an ambulance for him because we were concerned that he would vomit and choke. 'O' had no means of getting home so had to stay for the rest of the evening, stinking of sick. If it were me I would have walked home, rather that stand there all night smelling like that. In general door work is the same everywhere in the country. You are the barrier, which people have to pass to gain entrance to a club and you are also the one that can remove them from the premises, but you always get situations that you don't expect.

'CC' was one horrible doorman. Nobody liked him because he was a grass, shit-stirrer, thief, member of the tall story club and absolute knob-head, but his brother in law owned the club and employed him to do security along with us lot. He was not part of our team but we still had him in our faces all night. One of the things he used to do was go into the cloakroom and steal the girl's purses from their coats. He would then dispose of the purse and keep the cash. The worst thing about this was that for some of these girls it was their fare home, or if they had a travel pass it would get thrown away with the purse. He was one horrible bastard. It took quite a while to come to light, that it was him, and a few cloakroom staff were sacked for, wrongly, being accused of stealing purses. We grabbed hold of him one day just after he came from the cloakroom and forced him to empty his pockets. He had two purses on him which he claimed to have found on the cloakroom floor, and was returning to the manager. The purse thefts stopped after that. One day a young lad came to the door and asked if his girlfriend could pop in to use the toilet. There were no public toilets around and he said he would wait at the door with us and even leave a £10 deposit until she came back. 'CC' took the deposit and the girl was let in the club. The young lad asked me if I knew his brother, who was supposed to be a doorman over in East London. I hadn't heard of him, but there was no reason why I should have, there are a few thousand doormen in East London. I told him that his brother had probably never heard of me either. That's the way it is in any profession, you simply cannot know everybody. 'CC' butted in with,

'You know how many times we hear the line that somebody's brother is a doorman,'

It was true, we did hear that all the time but there was no need for 'CC' to be rude. He carried on with,

'Don't think you are gonna get free entry to this club by giving me that bullshit. You've got a fucking cheek. You can forget about your tenner and your girlfriend, now fuck off'

With that 'CC' walked off into the club. I said to the lad,

'Take no notice of that wanker, I will give you your tenner back and get it back off him later. I can't go and see him now because I've got to cover the door,'

The lad thanked me and shook my hand. He asked my name and said,

'There are not many nice blokes in this game like you, If you ever go to the ???? club, ask for Lenny, he's my brother and he will make you very welcome. He's the head doorman there,' I replied,

'Thanks, but I don't go clubbing, but I will keep it in mind. My name is Glen.'

I would never give anybody my real name because its so easy for someone to abuse your name at another club, like, *'I'm a mate of Glen's in the club over there. He said you would let us in'* This sort of thing was always going on in club-land.

Just then the girlfriend came out of the club crying that 'CC' had grabbed hold of her accusing her of trying to get in the club free and told her to fuck off before he gave her a slap. The young lad cuddled his girlfriend and said,

'That wankers dead' He then walked off still acknowledging me as being alright. That evening 'CC' was involved in a car road rage incident. A flask of boiling fluid was poured all over his face and head. It was believed to contain boiled water mixed with sugar causing the burning effect to stick like glue. It's an old prison trick that unfortunately found it's way to 'CC' that evening. If you're wondering where someone gets boiling water from in the middle of nowhere?

It's from such a thing as a disconnected radiator hose.

I obviously didn't see any of this but 'CCs' brother in law told us that 'C' was being treated in the burns unit for damage to his face and would be away for quite a while. He was too scared to involve the police. Maybe it was related to the incident with the lad at the club or maybe a pure coincidence, who knows?

If you work on the door and you are a bit of a tosser, you will get smelt out by your colleagues and also the punters. That will either be the last time you work there, or you will be the butt of all the jokes and ridicule associated with being a tosser. Long term - If one of the punters doesn't do you, the other doormen eventually will. Nobody wants a tosser on the firm, especially if they can't hold their hands up or they do something else that gives the rest of us a bad name. We need to look after each other, if a big fight does go off and one of the doormen is found hiding under the table, not getting involved, basically shitting themselves, they will lose all credibility and shortly after they will lose their teeth.

Someone who comes to mind here is 'Farting Martin.'

'FM' acquired this name because he practically lived on Tuna and Pasta, as part of his body building diet, and boy, did it make him fart.

'FM' was all shop window but had nothing in the stores. Meaning, he looked the bollocks as far a being a bouncer goes, but when it kicked off, he shit it like he was on laxative. Whenever it was about to go off he would find some way of avoiding getting involved. He would disappear to the toilet, or quickly disappear to the other side of the club making out that he was dealing with a situation. Sometimes it would be, the sudden emergency phone call that he just remembered he had to make to his dying Gran in Australia, or he would press the ringer test button on his mobile phone and make out he had a call and had to go to a different part of the club to get a better reception. He became so predictable that it was embarrassing. If we ever wanted a cup of tea from the night cafe all we had to do was talk amongst ourselves about how it was going to kick off any moment and also mention that we fancied a cup of tea. It was 100% guaranteed that 'FM' would volunteer straight away to disappear from the club for a while.

Although it was fun taking the piss out of him, every time he used one of his stories from his list of excuses, it also became embarrassing because some of the regulars in the club also began to notice that 'FM' was a little fight shy and were also taking the piss out of him. We had to do something about it, before people started to think that the other doormen were also like that and began to undermine our position.

Freddy was one of the other doormen who was a bit old school and just wanted to beat 'FM' up, because, he was pissed off carrying him. I convinced Freddy to let me have a chat to 'FM', to see if I could let him know what we were all seeing in him, that he was clucking like a chicken every time a problem arose. I was basically going to advise him to seek employment elsewhere, in a tactful way, so that he could leave with a little bit of dignity, rather than being carried out on a stretcher. It doesn't bother me if someone finds it uncomfortable dealing with confrontations. Being scared is a natural part of life, but I don't want someone doing a Houdini and disappearing on me when I'm about to get my face caved in by a group of lads.

I spoke to 'FM' and told him that we were tightening up security in the club for the next two weeks. We also agreed that we were going to leave our mobile phones in the office and would not make or receive any call within work times, plus would not leave the club to get snacks or go to the toilet if a situation was about to arise. I pretty much destroyed 'FM's list of excuses, without openly telling him he was a bottle job. The only place he could disappear to now was the storeroom where all the empty cardboard boxes were flattened and stored. We knew that he was using this room to hide because it would to stink of tuna farts, compliments of 'FM.'

The next weekend passed without any problem but before long we had a kick-off. 'FM' was nowhere to be seen. I had a plan to seek out a clucking chicken. When I was a kid, in the Naval Academy, I had this Japanese science and physics teacher who used to make the lesson fun by showing us little tricks and experiments. One of these tricks that have always stayed with me is the dry ice experiment. The teacher would put a block of dry ice into a can and put the lid on it. It was a push on lid like you would get on the Golden Syrup tins. After a few minutes the dry ice would evaporate and build up pressure inside the tin until it went bang!, and blew the lid off. I adapted this by putting dry ice into a can of paint, half filled. When it exploded the paint would go everywhere. As kids we would get the dry ice from any fishmongers but for some reason it became hard to get hold of.

I managed to get this block through a mate of mine who was an actor. It is still available in the theatrical world where they use it for special effects like graveyard scenes and things like that.

Although we knew that 'FM' was hiding in the store room, we were all too busy elsewhere to actually see him go in there, so I set up a gallon can half filled with paint and put a block of dry ice in it and banged the lid shut. We quickly created a make out situation leading 'FM' to believe that it was going to kick off any minute, without fail he disappeared. It wasn't too long before the can exploded but what we didn't expect was that 'FM' wasn't in the storeroom because he was fight shy as we all thought. He was in there having his dick sucked by one of the bar staff. But even more revealing it was poofy Wayne as we called him. Well! What a fucking surprise that was for us. Poofy Wayne was on his knees going down on 'FM' as the can exploded. Both of them had their trousers around their ankles and the paint just covered them. They were absolutely smothered in red paint. As the can exploded they both shit themselves and raced out of the storeroom directly into view of us lot who were waiting to see 'FM' get caught out. Obviously we hadn't counted on his bum chum being in there as well. Needless to say, neither of them came back to work at that venue. We never had a problem at that club anymore with doormen disappearing. They were too worried that they may get accused of going to the storeroom for a blowjob with one of the gay members of staff. I must admit that the gay thing within the world of doormen is a pretty rare thing.

I have worked with a few strange doormen though and I can remember one night that was very freaky for me.

I was working with one bloke,'L', who in my eyes was a bit of a lowlife. He would blow all his wages on drugs, gambling and stay out all night shagging anything that wanted free entry to the club. All this while he had a wife and young baby at home with no food in the cupboards. On a few occasions his wife would phone me and ask me to try and get some money from him so that she could get some food in, or put some money in the electric meter to keep the house warm. Although I felt sorry for her situation, she was an adult and could sort her own life out, but when it comes to kids, I'm a soft touch.

On a few occasions I sent bags of shopping around to her, anonymously, in a cab and made sure it had plenty of baby stuff in it.

She knew that the shopping was something to do with me and probably thought that I had managed to get some money from him to sort it out, but I didn't. I didn't tell anyone else that I was sending the food round because I didn't want anyone going,

'Look at Jamie, the fucking charity man,' or,

'Yeah we all know what's going on there!'

Because my reasons were completely innocent. I simply had plenty of money at the time and £30 spent on a bit of shopping didn't make the slightest dent in my pocket. As I said before, when it comes to kids suffering in any way, I'm a soft touch.

Every time I sent a bit of shopping round to the house I would tell 'L' (Lowlife) that I did it for him on his behalf and that he could settle up with me later, but he never did. It did piss me off, but to contain myself I had to keep reminding myself that he hadn't asked me to do it. It was my decision to do it, because, it made me feel a better person. The thing that angered me most though was the fact that he was shagging a different girl every night and on the odd occasion that he bothered to go home to sleep with his wife, he would have sex with her, exposing her to the dangers of Venereal diseases, Hepatitis, Aids and so on. He was an irresponsible, sad lowlife.

Anyway we had finished work that night and he had promised to drop me home. Normally I would drive myself home but this particular weekend I had arranged to meet an old pal of mine at a pub in South

London which was not far from where 'L' was living, so he said he would drop me off on route. I did not want to drive that night because I wanted to chill out after work with a few beers.

Work had finished, we had cleared everyone out and I was chilling out with a beer whilst waiting for 'L.' He was in the rear of the club doing his thing with a couple of tarts he had pulled earlier. They were either shagging or doing drugs. I didn't get involved, I just preferred to chill out on my own, reflecting on the night's events. It was one of the ways I would unwind. About half an hour later 'L' came to the bar and said that he was ready to drop me off but had been invited to a party, by a girl he was just with, and that it was not far from the club, so wanted to pop in for five minutes to say hello to someone. He said that I could go in with him or stay in the car. I decided to wait in the car. We got to the house and 'L' went in saying that he would be out in a few minutes. Twenty minutes had passed and I was getting pissed off, I was also dying to go to the toilet from the couple of pints that I had just downed at the club. I went over to the house and introduced myself to the guy on the door who was playing bouncer and he let me in. I went straight to the toilet and then went to find 'L.' All the lights were dimmed and it was very dark. Music was playing and it just seemed like your average party. Couples scattered around chatting, kissing and getting it together. I heard 'L's voice, he was bit of a loud character who wouldn't go unnoticed anywhere. He was snogging this girl whilst having his hand up her skirt, fondling her. I thought what a lovely girl, if only her parents could see her now. I said,

'L', come on, I'm fucking pissed off waiting for you, drop me off at the drinking club to meet my mate and you can come back' With that the girl turned around to look at me and said,

'Fuck off will you, can't you see we're busy'

It was the husky voice of a man. 'L' was snogging and touching up a fucking pre-op transsexual. I just looked in astonishment and said,

'What the fuck!'

I just didn't know how to react. 'L' said,

'Jaym, I can explain, listen'

He then pushed the Tranny off his lap and stood up. The Tranny started shouting and the big guy on the door switched the lights on.

The whole room was full of blokes with Trannys. It was a fucking Tranny brothel. I said,

'Fucking hell, I'm out of here,'

Then left pretty lively. The guy on the door was a girl and looked as if she was gonna punch fuck out of me. I don't care what consenting adults do as long as kids or animals are not involved and nobody is forced to do anything they don't want to, I just got freaked out by 'L' being married but also liked to suck dick?

Anyway 'L' came running out to the car and got in,

'Jaym, the one I was with was a girl, it weren't a geezer, honest, what do you take me for, you see the amount of tarts I get'

I just sat there freaked out but also laughing. All I kept repeating was

'Fucking hell'

Shaking my head in astonishment. He dropped me off and went home to his wife.

He stopped working with me after that and I was certainly glad to see the back of him.

So apart from the strange and freaky events, generally all door work is pretty much the same but I will share a few more of my door experiences with you, to show its not all glitz and glamour, just as an insight to show you what it is really like.

My aim is not to pick out every fight that I have won in order to become your hero. To be honest with you, the worst fights and most dangerous encounters I've had are not even in this book. I just want you to see how I, as an individual, handled different situations, and tried to avoid the ugliness of hurting people which is the total opposite of what I wanted in my youth. I no longer wanted to prove my physical ability. I had already served my apprenticeship and passed that test. I was now trying to find an easier route through life. I suppose I was trying to make more use of my brain than my brawn. I was so confident in my own ability in being able to look after myself that I had nothing to prove to anyone and made sure that I didn't fall into this ignorant small minded trap. I felt I had now reached a level of maturity with fighting that could be summed up by the famous Bruce Lee quote *'The art of fighting, without fighting'*, or so I thought!

I had been working at the 'Snap club' for about a year and needed a change so took on a position at the 'Aslas club' just up the road.

The 'Aslas club' was a strange and new experience for me. I had been in the game for about 15 years but had never worked a club like this before. No one spoke English and I only Speak English, so this was like a new world.

I was working with one other doorman, 'NW', who had just arrived. He pulled up just after me and asked if I had eaten yet because he was going to have some grub at the 'Aslas.'

I wasn't that hungry, but I knew that I wasn't going to get my 1 o'clock meal, that we normally got at the Snap club, so I decided to have something down at the Aslas club. It didn't take much to convince me, the food was free and I heard very enjoyable. We set off for the Aslas club, which was only a couple of turnings away from the Snap.

On the way up there 'NW' told me he had only actually been working the door there about a month and had a fight only the weekend before with a giant of a man who got caught nicking purses from handbags. 'NW' said that the guy was giving him a bit of a pasting and was trying to throw him over the balcony, so he squirted him with some CS gas to put a stop to him. He then knocked him out.

CS gas will temporarily blind you and make you short of breath and panicky. The gear that comes in from France is about 85% potent compared to the CS the police used in their trials, which was only 5% potent. The police even stopped using the 5% stuff, because it was so lethal, so you can imagine how bad the stuff from France was. So after he had sprayed the big guy with the gas, 'NW' took advantage of the situation. He didn't say how he knocked him out but the bloke was still out cold when the police arrived, or he was making out to be, so he must have taken a bit of a whack.

'NW' explained that the Aslas club was a Latin-American restaurant and nightclub that played Spanish-Brazilian and Mexican music.

It draws in punters from all these different cultures and others who were interested in this scene. So actual English speaking Londoners didn't really go there a lot. In fact hardly anyone there spoke much English. The club must have catered for about 370 customers and out of that only about 20 were English. The main concern I had about working there was that the punters were not searched, because the management said that they didn't want to cause offence to the paying public by doing so. That's what they said anyway! Personally, I believed, that, it was because they were taking a £5 from everybody that came in the door, and the quicker they got them in, the more money they would make. Everyone was paying cash so someone was having a nice little earner there but it certainly wasn't us. The £5 didn't include any food or drink. It was simply an entrance fee.

So really at the end of the day, the two of us were in the uncomfortable position of being slightly out-numbered by gangs of Latin Americans, who are not unknown for being a knife carrying culture. What a poxy way to earn a living! But living on the edge of danger sort of gave me a buzz at that time in my life. I didn't have much else happening in my life at the time so I suppose I needed some excitement to stimulate my brain. Loneliness can also be a real killer, especially when the dreaded depression sets in, so doorwork is a good therapy for that. I was always, sort of, on a high because I have always suffered slightly from anxiety, so it made me feel like I'd had downed a couple of pints, without drinking, and I was getting paid to do it, quite good really!

The thought of 370 people being able to tear two doormen apart with their bare hands was bad enough but when we got there I soon realised that one of us had to do the door and the other had to walk the club downstairs. So in reality we were each on our own.

We didn't even have radio mikes (walkie-talkie's) and 'NW' didn't have his gas anymore, due to the management's request. Obviously he'd gassed somebody a few weeks earlier and they were worried about a police raid happening and him getting caught with his gas on him and getting done for a firearms offence.

So neither of us carried any form of weapon, it was down to whatever we were capable of doing. The management were so worried about us carrying anything weapon like, that they would make us leave our jackets and pocket contents in their office. It was bollocks that the two doormen would practically get searched; yet we were not allowed to search the punters. So neither of us carried any form of recognisable weapon. All I carried was my trusty Mini maglite (Pocket torch) and twenty years of experience in the fighting arts, 'NW' carried a telescopic baton, which he justified by calling it a search probe, to protect him from needle stick injuries. Apart from that all we were relying on was our skill, training, tact and luck if it went off.

We both obviously needed the money; otherwise we wouldn't have been there (well I certainly wouldn't have been).

When we got to the club it was pretty quiet, only the restaurant section was open so we ordered our grub. 'NW' had what seemed like a 20 course meal and he packed it all away where I was content to have a plate of potato skins, filled with cream cheese and spring onions. It was unusual but very nice. I have always been limited for choice being the only vegetarian doorman in the world.

Anyway after eating we put on our red jackets that were supplied by the Aslas. They were a bit like the old Happy Days TV programme, high school jackets. They had 'Aslas' plastered across the back of them, and then we were ready for work. I felt a right prick wearing this jacket but was told it was to take the aggressive image away from the door supervisor, making them more approachable to the members of the public. It certainly did that. It made us look like a pair of fucking clowns. It also made us walking targets; you could spot us a mile away.

We each took turns on the door in twenty-minute blocks, so one was on the door for twenty minutes while the other was downstairs and then we would swap over. Working the Aslas club was a pleasant atmospheric change from the House and Indi music that I was used to at the other clubs. You could easily forget that you were even in a nightclub in the heart of London. It was as if you were in a local, back street, dance festival in Spain or Brazil. The atmosphere was electric, I could see how easy it was to become too relaxed, switch off and to get in the mode of the atmosphere, but it was guaranteed that as soon as you did, some clown would try it on with you. I always tried my hardest not to drop my guard. I didn't want to get caught out and end up getting sorted out by one of the more macho types that frequented the club and wanted to make a name for themselves.

We used little pocket clickers to count the people in and out of the club, as part of the fire regulations. If a fire did break out and the Fire Brigade were called, or any of the local licensing officers turned up, we were able to tell them exactly how many people were in the club.

It was horrible carrying the clickers, because, every time you clicked that button, the odds against you got higher and higher.

At most times there were five hundred people in the place, the limit of 370 was a joke, and what we're we going to do, two people against five hundred. It was ridiculous! But the management just went on and on taking them in.

One-night things were going well and everyone seemed content to get drunk and have fun, except for one guy. You always get one. I could see 'NW' having words with this guy, who was dressed up like a Mafia boss. He had the braces, the clean-cut suit and he was strolling around like he owned the place. In the way he was acting, he obviously thought he had a position of power within his little pond, that being the club. I had eye contact with 'NW' to let him know that I was ready to go for it but he returned the nod that all was ok. The situation had seemed to diffuse and that was it, or so I thought.

About half an hour later a drunken friend of his came upstairs to collect his overcoat and waistcoat, which indicated to me that he was ready to leave. It was about 1 o'clock in the morning and a group of three ladies were just collecting their jackets from the cloakroom at the same time so that they could leave. Our drunken friend decided that he didn't want to wait for the ladies, to collect their belongings, and jumped the queue. The manager whispered to me just to let him leave, he didn't want any hassle and just wanted him to go peacefully. I wasn't too happy with this. I felt he had overstepped the mark and I prefer to nip things in the bud before they get out of hand, but I was the worker and not the boss so I had to follow orders.

Without provocation, our drunk had become offensive to the ladies, calling them 'Fucking Whores' and continued, screaming,

'All women are sluts and whores'

This rang a bell in my head, which seemed to be saying, 'Throw this prick out.' I hate derogatory comments like this towards females.

I could have just whacked him but it was a bit hard really as the door camera was linked up to a video security machine, so I just asked him politely to leave the establishment. I said,

'Listen to me you fucking fat cunt. Get your arse out of these doors now. Otherwise I am going to knock you senseless' He replied,

'Soy Capitan'

Along with some other crap I didn't understand.

He was trying to tell me that he didn't speak English. He came out of his mother tongue to say

'I do not speak English, my friend'

I knew he was mugging me off as I had seen him speaking to 'NW' earlier in the evening and I know 'NW' only speaks two languages, ones cockney and the others bullshit.

So our drunken friend was a liar, making out he couldn't speak English. I really wanted him to turn on me, as I knew the camera was on and I would rather be caught on video defending myself, than sticking one on him without provocation. I didn't want to be caught on video appearing to use one of the general public as a punch bag, the last thing I wanted was to get nicked and also I didn't want to lose my door working license over something as trivial as this. Our friend tried to use the language barrier with me again but that didn't cut any ice. I speak another language, which helps me deal with this type of situation. Its called 'Body Language.' So I poked him in the chest with my left hand while switching on my right. By switching it on, I mean preparing it ready to fire, preparing to knock somebody out. I had it ready just in case I needed to use it. Then I pointed to the door. This means the same in any language. Body Language is a good universal language that anybody can understand. This was my interpretation of 'You-out now!'

He must have understood body language because he said,

'Yes, my friend, I will go now'

I was not going to let his passive, apologetic response work with me. I was staying switched on until he moved. As he walked out, he turned and bad-mouthed the ladies again.

I didn't understand what he said but they obviously did to begin shouting back. Outside the club was where I wanted him, because it was out of the range of the cameras. At this time there was about 70 punters lining up outside the 'Tent club', which was the venue across the road facing us. They were a mixture of punks and skinheads. None of them would have gone witness if I had to introduce my friend to the pavement with a right cross.

Now that I had matey outside I didn't give a shit whether it went off or not.

My job as a doorman was done. But being a member of the human race I had to make sure he didn't hang about to pick up the girls or pick on them as they left. Now there was just the two of us and he was outside of his pond where he was the big fish, his cockiness and aggression had gone and he was trying to get me on board as his friend and hopefully regain entry to the club. I just told him,

'Fuck Off and get out of my sight'

Which he understood quite clearly. That ended that.

As I was explaining the route to a local restaurant to a tourist, who asked me if I could help. 'NW' stuck his head out of the door and shouted,

'You dirty bastard'

He wasn't name-calling me, but rather the drunk who I had just escorted out. He shouted again to me,

'Look, look what he's doing'

I turned round and the drunk man was pissing on the overcoat of the stranger I was giving directions to. I could do nothing but burst out laughing. I'd never seen anything like it before. I couldn't do anything but turn and walk back into the club. I didn't know what to do. I thought I'd let the pair of them sort it out and squabble between themselves. Nothing came of it and they both went their own ways. But I learnt a very important lesson from this situation. I should have stepped back and kept eye contact with the drunk I escorted out. Next time I stupidly switch off and lose eye contact with a problem customer, it may be a bit more than piss that I have to worry about. It could be a bottle in the face of a knife in the back. So you can see here what I was talking about, when I said it was easy to lapse and switch off for a while. I learnt a great deal from this comical experience and took it as a wake-up call to stay switched on.

Every night wasn't all bad at the 'Aslas.' Generally it was quite a calm place but sometimes it used to go off, time and time again. I remember working down there another time; this particular night there was a lot of Brazilians in. They had a live Brazilian band in playing music. The T.V. screens in the club had a videotape of Brazil going through and winning the World Cup. The tape was played every single night without fail. It was either that or cartoons of 'Speedy Gonzales.' The punters seemed to love all that. I suppose it must have reminded them of home, especially the football. It's ok. if you like football, but football is not for me, especially the violence normally associated with the game.

It was about 10 o'clock in the evening and I was working on the door controlling the queue. The place was packed and we started a queue about 9 o'clock on a system of one in, one out. The 'Aslas' was at its maximum numbers that night and we needed to keep a tight control on it. There is always a lot of hostility in a crowded line or queue because the punters just want to get inside the club, get out of the cold or the rain, and start to dance, drink and have a good time.

I was the barrier from stopping them doing that and not usually their favourite person. They didn't give a toss about fire regulations or the maximum numbers allowed in the club at any one time. All they want to do is get inside. I always tried to be pleasant and reassuring by telling them that they will not have to wait too long, but how can you be reassuring to a hundred punters who are out in the freezing cold or waiting in the pouring rain, after getting all dressed up, knowing that they may not get in after waiting for ages.

'NW' was downstairs in the warmth of the club and was supposed to have bought me up a coffee, we are not allowed to drink alcohol on the door. The coffee's ok. and that suited me. I would never drink and work at the same time, and I always had to drive home after the night's shift. I was pretty cold myself, standing on the door all night so I needed a bit of warming up.

I looked down the stairs to see if my coffee was coming and saw 'NW' trying to break up a fight between the Algerians and the Brazilians. It looked like it was diffused until one of the Brazilians pushed 'NW'. This seemed like a command for both sides to do the bouncer. All I could see was them steam into him. I flew down the stairs and into what looked like a rugby scrum and in I went! Its times like this that you really do feel your adrenaline giving you the strength to steam in, your courage says get the fuck out of here, but as hard as it is, you are paid to get in there and try sort it out.

We managed to separate the two sides and dragged the two ringleaders up the stairs. A couple of the guys were still on the floor, bent over from injuries. At the top of the stairs the Algerian had calmed down. It was now shut up or 'swallow my fist' time because I

138

was just pissed off by now. He was explaining his side of the story to me while 'NW' was trying to get some sense out of the drunken, furious Brazilian. The Brazilian was asked to calm down but he just got louder and louder. The manager asked him to leave and said he would be able to come back the next day, when he had calmed down a bit, but the Brazilian was having none of it. He shouted something in Brazilian to his comrades downstairs, which we, obviously couldn't understand. Whatever he said, suddenly, prompted, all his mates to come tearing up the stairs, coming for us.

The manager was grabbed hold of by the Brazilian and got into a headlock. 'NW' immediately jumped on to the Brazilian and tried to drag him off but by this time we had about fifteen Brazilians on us. A few good centreline bodyshots and headbutts evened things up a little bit and they decided not to carry on. 'NW' was still hanging on to the Brazilian and the Brazilian was still hanging on to the manager. The Brazilian let go of the manager, which was a good decision on his part as 'NW', was in a position to snap his neck. The Brazilian boys grabbed their man and dragged him out of the club. They were still kicking and screaming at the manager at this time but we managed to get them out of the doors and that was luckily was the end of that.

They were banned from the club so we expected it to go off again, you know like a revenge attack later on that night. We were on edge all night long waiting for something to happen. We didn't know if they were going to turn up in a car and steam us, so all we could do was wait.

While all this was going on, the queue still hadn't budged. It always amazes me that punters who see things like this happening still want to pay a fiver to get in the club!

The first few people in the queue tend to try and make eye-contact with you every time you walk in and out the door in the hope that you are going to let them in. But as I looked at the queue this night the people at the front had changed since I had last been out before the beginning of the fight. I new that we had not let anyone in and that nobody had left the queue but something had changed. Yes, that was it-we had a queue jumper! It was a drunk we had thrown out and he was intimidating the ladies in front of their boyfriends. He may have gone unnoticed except that he had the longest set of dreadlocks on a non-Rastafarian I had ever seen. How did I know our man wasn't a Rasta? Well I spent quite a few years living in Tottenham, N17, where the 'Broadwater Farm riots' went off. Being a member of the community you learn the genuine article from just a fan of the Rasta way, just like you get to know a fighter from a talker. You just get to know! Anyway it mattered not to me whether our man was a Rasta. All I cared about was that he had jumped the queue and was pissing people off. I asked him to join the end of the queue and he replied,

'I've been here all night man. I'm not moving'

I didn't need this sort of drama after the fight we had just had. I could have done with a break for a little while but in this game you just don't get a choice. When a problem crops up-it just happens. A bit like sneezing really!

I just said to the guy,

'Look! Move now or you won't come in at all, Ok?'

But he wasn't interested. He retorted,

'Who's gonna stop me man-you?'

Bang! That was it. I hit him, dead centre-line just where the two halves of the rib cage meet, with a vertical short jab. His legs gave way and he dropped to the floor.

I knew none of the punters were going to complain about me in case they didn't get in, and also because I was doing them a favour-he had jumped the queue.

The manager must have seen the guy drop because he came running out. I thought I was going to get a bollocking. I was lucky; he didn't see what happened. Although he wouldn't have wanted the guy jumping the queue, he certainly wouldn't have wanted me bashing him one in front of all the punters. The guy obviously was allergic to being hit because he was on the floor being sick all over the place, and all over himself. So with the manager's help we dragged him down the road and left him there. The bar staff came out and washed the sick away. The night was not turning out to be one of the best but it could have been a lot worse.

I had one more incident that evening where I stopped a guy slapping his girlfriend. He told me to 'Fuck off' and grabbed my jacket to display his toughness in front of his mates. I had my mini maglite torch in my hand at the time and the butt of it somehow found its way into his chest-plate. He screamed like fuck, he thought he had been stabbed, and ran up the stairs screaming,

'I've been stabbed, I've been stabbed'

He ran out of the door and that was the last I saw of him. I never saw him again. I was sure glad to get home that night.

As I've said before though, not every night on the door was all action. Sometimes, it was pretty boring. In general though there was always some kind of incident going on but most of them were pretty minor. You had to stay switched on all the time, just in case it went off.

The queue started earlier on this other particular day because the restaurant section of the club was well booked and we had to hold back the numbers to allow for those who were coming in to join tables. The crowd were getting agitated and uptight as it was getting nearer to 9 o'clock, because they knew that entry to the club was free before nine but was a fiver after. They thought it was a scam set up by the management or us to charge them money but it wasn't. We just had to wait for the tables to be cleared until we could let any more people in. I was on the front door taking all the shit this night as usual. I was still trying to smile and be pleasant but I wasn't flavour of the month with the crowd. Some big fat bastard in a suit and tie decided he wanted to come into the club and was not going to join the queue. He obviously mistook me for the invisible man because he tried to walk straight through me. I politely refused the prick entry, which he found offensive. Through his drunkenness he said,

'Take your best shot. I love fighting'

I just replied quite calmly and within hearing range of the queue,

'I'm very sorry that you misunderstood me but I will explain once again. It will cost you five pounds to get in but you will have to queue like everybody else'

Now I could just have refused him entry, due to him being drunk, but I thought that if he was to join the queue, he would be there for about an hour before getting in and would just get pissed off in the end and go home. The fat drunkard mistook my politeness, in front of all the witnesses in the crowd, for being a sign of weakness or me being scared of him. You know, like me backing down after he had been aggressive to me. So he decided to have another go.

'Let me in the door now otherwise it is going to get ugly', he said. I

141

thought, that's it you bastard, I'm gonna lay you out. With that I said,
'Let me speak to the manager and I'll be back in a second'
I could see that he felt very macho but I felt like biting out his carotid
artery. You have to do things in a clever way when there are 70 or
100 witnesses in front of you and again the last thing I wanted was to
get nicked and lose my job. I was there for the money and I certainly
did need the money.

The manager was already at the top of the stairs because he had seen
me on the video camera and could tell that there was a problem at the
door. 'NW', the other doorman, was with him. I said to the manager,
*'Get 'NW' to cover the door for me because this wanker wants to
fight me so I'll have to take him round the corner and do him'*
The manager knew the score and said,
'Please escort him away in a pleasant manner'
In other words *'Hit him as hard as you want but make sure he
doesn't return - I have no knowledge of this if anybody asks me'*
This was fine by me. If I didn't need the money, I would have nutted
him on the door at the beginning but obviously I would have lost my
job and been nicked for assault. 'NW' smiled at me as I walked out.
He knew how I worked. If I walked somebody away from the main
door, he knew it was time for me to deal with them. I came back out
to the fat boy and said,
*'I have spoken to the manager and if you come over here I will
explain all to you'*
He felt very proud of himself and was swinging his shoulders as he
walked over to me. He obviously thought he was going to gain entry
to the club and that I was actually scared of him. When we got round
the corner and out of the sight of the crowd and the video camera, I
was switched on and ready for combat.
*'Listen to me, you piece of shit. I have just clocked out of work so
I'll take you up on your offer of a fight. I'm gonna tear your
fucking face off, rip off your head, then shove it up your fat arse.
So fuck off out of my face or try your luck now'* I said.
Now, when I switch into combat mode, the more verbally aggressive
I become and I just wanted to kill this piece of shit that had wound
me up. The fat boy shit himself. His face looked like death and he
just wanted to be elsewhere. After his first few apologises, he

became boring. His scared stuttering was pathetic. His adrenaline must have tore his bottle to shreds. He now wanted to shake my hand and be friends. I wasn't having any of it; I wasn't having any of that bollocks.

'Get out of my fucking face now'

He went, still apologising to me as he skulked off. I then had to unwind and switch back to being smiley and pleasant to go back to work on the door.

I much prefer dealing with this type of situation verbally than using violence. It was a skill, which I wanted to try to perfect. If I was not able to do this, I would use my fists anyway but that would always be my last resort.

It's funny though. Trouble in nightclubs is not always from those who you'd expect it to be. Because in the Aslas club there was a restaurant as well, you would get different groups booking party celebrations, wedding celebrations, hen nights, you know whatever, all within the Latin-American theme.

I can remember one time, there were two tables booked, about fifteen people on one and about twelve on the other. Throughout the night somebody decided, for a laugh, they would throw a piece of their food across the floor, which whacked somebody in the head from another table. Although when you are drunk that may seem funny, like a kid at the pictures throwing popcorn at people, but when you're on the other end, receiving a greasy potato skin in the back of your neck, pisses you off. Now, I didn't see this actually happen. I was upstairs on the door and 'NW' was downstairs but one thing lead to another and they all started fighting and throwing food. The waitress got caught up in things, somewhere along the line, and went up to the manager who decided he wanted them all out. 'NW' asked them to leave. You can imagine the situation where there were about 25 adults who had probably spent about £30-40 each a head on food, had just eaten their food and had only just paid for it then being asked to leave the club.

They were pretty pissed off but they had brought it on themselves. Anyway 'NW' asked them to leave. Now, 'NW' was not quite as diplomatic as I was. If I wanted someone to leave I'd ask them politely but firmly so that they understood what the score was and that I was switched on to be aggressive, if need be. With 'NW', he is just aggressive and that's it. He had no communication skills.

I think it is down to immaturity and his age plus not having enough experience working the door, communicating with people.

Anyway 'NW' asked them to leave. One of the punters said to him, *'Fuck off. You're too short to tell us what to do'*

'NW' looked quite small but he said he was 5'9'. I'm only 5'9' but 'NW' looked about 5'5' to me and I felt that I was a bit short to be a doorman. Anyway 'NW' came up to me; obviously he had never had to deal with a situation like this before. He said,

'Jamie, I've got trouble with some guys downstairs. There's about twenty-five of them and they won't leave'

'NW' just stood there. He didn't ask me to deal with it but I understood that's what he wanted.

I went downstairs and politely asked them to leave. The headman of the table came over and told me that they had paid for the meal and were not leaving. I said,

'Look if you don't leave I'm gonna drag you up them stairs in front of your mates and if you resist I'll bash your fucking head in'

With that he quickly organised his party and they decided they were going to leave. I had to do a similar thing to the other crowd because they were cheering as I threw the first table out. So at this time we had about twenty-five people at hand, about fifteen guys and the rest girls, and fifteen guys could have torn me apart, trained or untrained it wouldn't have made a difference. There would have been nothing I could have done against all fifteen of them, but luckily enough after becoming switched on and semi-aggressive, I had managed to persuade them all to go and thankfully that was the end of it. They all went.

So again I had to use my passive side to win this challenge. If I had knocked someone out, it would have been a no-win situation for me. Possibly if I had hit one of the guys, all the others may have backed down but then again I could have been faced with fourteen others ready to then rip my head off. I didn't want to take this chance. So I was glad I was able to psyche him out. If one of them had gone for it, I would have had to knock him out anyway. That's just one of the downfalls in working the door. Anyway that situation ended fine and

I soon forgot all about it.

The next night something very similar happened. We had about seventeen people come in, who I think were Greek, and book a meal plus there were four youngsters in a separate group. I'd say kids but obviously they were over eighteen to be allowed in the club area of the licensed premises. I think they must have been about nineteen or twenty and there were two girls and two boys and they were also eating in the restaurant. I was actually working downstairs at this time and 'NW' was working upstairs. As I was coming down the stairs I saw this big mob of Greek's bashing the shit out of one of the young guys. Straight away I dived in, separated them, took a whack from somebody, I don't know who but I got one in the eye, and managed to break them up. I couldn't get to the bottom of what the problem was but obviously the young fellows weren't going to start on seventeen big Greek's and these Greek guys were BIG lumps. Now the problem we had here was that they had been brought down to the club by coach and the coach was not going to be there for another hour. Now, if we threw them out, we would have real trouble on our hands because they would just hang about and we had a queue outside the door at this time and didn't want any trouble started outside. Luckily enough one of the four youngsters who was involved in the fight, asked to be escorted up the stairs and down the road, he was pretty bashed up.

They were a bit worried in case the Greek's were going to jump them outside the club. So I did that, I took them upstairs. I then told the Greek guys to calm down otherwise they were going to be out and they were quite ok about it all. In fact the top man was quite embarrassed that it had happened, he wasn't really sure what had gone on and promised me that he would keep his crowd under control until the coach came and that was it. They turned out to be real sound guys who unfortunately had an arsehole with them.

Again for two nights running, I was in the situation where I was left on my own against the mob. Even with all my training and street knowledge, I knew that the only way I could have dealt with that many people was to use a passive approach and try to psyche them out. Aggression would have got me hammered. I felt very pleased with myself. I had managed to talk my way out of a fight situation and it was more training in what I wanted to achieve in the art of word play and body language rather than getting into a fight. I knew I was capable of fighting, I could fight and I knew that I could come out on top against most people, but this was not the way I wanted to play things. It would only have taken me to get nicked once and I would have been out of work. So it is very, very hard. As a doorman your hands are tied. It was quite strange really, because I knew from previous fights and situations we have had on the door, especially with the Brazilian who had grabbed hold of the manager, that 'NW' would always steam in there. He wasn't scared of getting in a ruck, but I could tell that something was changing with 'NW' and I couldn't understand him.

The last couple of weekends, he had seemed to be shying away from the aggravation side of the trouble and I knew it wasn't any sort of clever move. He wasn't educated in the thinking way as I was in the art of combat and fighting skills. There seemed to be some other reason that he was becoming a bit confrontation shy, and I couldn't work it out. The weekend after that, a lot came to light.

The following weekend I met 'NW' down the road to the Aslas club. We had to meet our boss who would tell us which club we would be going to work at that night. I had been down the Aslas club quite a lot recently because it has been kicking off and the boss knew I was pretty handy if push come to shove, but also that I could diffuse situations without taking liberties with people. He also wanted me to try and train and educate 'NW' into being a mature good doorman but that is not an easy task if someone thinks they have all the answers. So because of all these factors I used to end up working down there quite a lot. He paired me up with 'NW' again which was beginning to concern me because I was having some reservations about his reliability when it kicked off. As we were walking down to the Aslas, 'NW' was having a fag, which he often did before going into the club, as we weren't allowed to smoke while we were working, although I don't smoke anyway, but as I looked across at 'NW' I noticed that he was smoking a joint, and that really did piss me off.

For one, I am totally anti-drugs of any form and one of the drugs that really does piss me off is dope because, unlike any other drugs which people take either orally or inject themselves, which affects them and nobody else, dope is different. The trouble with smoking dope is that if you are in that environment where people are smoking, you become a victim of passive smoking and I certainly did not want to be breathing that in.

So we were walking down the road and 'NW' is smoking this joint, I didn't say anything to him about it, instead I turned to him and said,

'What have you been up to today?'

'Oh, I've been doing shop retail security all day', he said.

'What after you had been working all last night?' I replied.

We had done the Aslas club the previous night until the early hours of the morning.

'Yeah' he said.

'I'm knackered. We finished work at three, I was at home at four and had to get back up at eight and I've been on my feet all day working in the shop down town doing the shops security'

I told him that he must be mad working during the day as well.

He told me that the money was good. I said,

'Yeah, the money may be good but I couldn't do that. Work all-night, then all-day and then all night again. You've got to be alert in this job'

That was the end of that conversation. I knew that he was trying to put across to me that he might not be so switched on that night, and that he was a bit tired. He didn't actually say that but I knew he was trying to forewarn me. What I wasn't happy about was that he was smoking dope before work

What he does in his own time, after work, was up to him, but you know, there was only me and him against anything between 350 to 500 punters and if it went off in a big way he's the guy I've got watching my back. If he's out of his nut, shouting, giggling, having a good time, that doesn't help me much if I'm getting a knife in the spine. So I was concerned. I was going to have to deal this.

There was nothing I could do then though because we were both sent to work straight away and I had the choice of working with him or working on my own completely. He was better than nothing.

Anyway, it was quite strange because throughout the night 'NW' started to get very loud. The affects were that he was talking a lot and

147

very fast and he was showing the signs that he had been on some kind of Amphetamine, like Sulphate, some sort of Speed, because he was really, like, bouncing about. It was the total opposite response than you would have smoking dope. Which makes me think now that perhaps it was something else he was smoking, I don't know, but it smelt like dope to me. Perhaps he had smoked a bit of dope and then had taken something else to liven himself. I don't know what happened. All I know is that he was buzzing about like mad, which you don't do on dope. Although I have never ever taken any form of drug in my life, apart from headache tablets, I do know the affect they have on people because I have had it around me throughout my life through friends and family plus I do understand the drug scene. So I know from the way he was acting, it was not dope he had taken, or not only dope he had taken, but something else as well. I just don't know what he was on.

Anyway throughout the night he was acting totally out of character. He was being very loud and talking absolute bollocks to everybody. Everybody was his mate, everybody was his friend. He just let his barrier down. He totally destroyed the image of the doorman and I was really concerned about this. He was showing a sign of weakness and I could see that people were slowly taking liberties with him. They were saying,

'Come on mate, here's our fiver to get in. Go on let us in!'

'NW' was saying,

'Oh, go on then', and letting them all in.

We didn't benefit by it; we didn't get the fivers. So I really didn't know what he was playing at. All I can imagine, well I know, is that he was out of his head on something.

Anyway, that night a few incidents went off. A few fights, which I had to deal with and 'NW' was just never there. I was getting really concerned. Then I got in a fight with one guy who I actually had to bang out. I hit him on the chin and he went down and he had to be escorted out of the club. Now, 'NW' should have been there and he wasn't. I said to the manager,

'It's going off quite a lot tonight. Can you get 'NW', get him over here, to help me deal with a situation on the door?'

The manager went down to find 'NW'. He was gone for almost ten minutes, which was a long time when you are dealing with a situation, and he still couldn't find him. Anyway, I was really concerned and pissed off working with 'NW' because he just wasn't normal. As the night went on, I saw the manager speak to 'NW' and say to him,

'What are you on?'

'NW' just laughed and said,

'What are you talking about?'

The manager said,

'Come on, what are you on? You're just acting differently tonight'

'Oh, leave it out, leave it out' he replied.

He was acting totally out of character. 'NW' came over to me.

I have a longstanding injury and problem with my nose because I've got a busted nose from year's back. I've had my nose broken too many times and it's really affected the nerves inside creating a habit of sniffing.

'NW' walked over to me, as I was sniffing, and said to me,

'What are you on? You on coke?'

'What the fuck are you talking about?' I replied,

'You're on coke. The lining of your nostrils have gone and you're always sniffing. I know I've seen it. What have you done, a couple of lines?', he said, It didn't occur to him that I have had my nose busted dozens of times in the hundreds of fights I've had. I was getting really pissed off now.

'Look, N, what are you on about?'

'You're on something', he said,

'Listen, I don't even drink alcohol when I'm working. I don't smoke cigarettes and I don't take drugs. So cut that shit out now' I replied.

He came back with,

'Yeah, leave it out, leave it out. Come on, what you on?'

He was completely gone. He didn't know what was going on, you know, he just wasn't with it.

He was supposed to be protecting everybody in that club, protecting the management and protecting my back as well. So, I just couldn't take it anymore. The manager then made another complaint about him to me, not officially, but you know, through the ranks. He said,

'NW''s getting a bit lazy you know. Something's not right. You're

going to have to deal with this otherwise we are going to have to
pick another team to work for us'

Well I didn't want to lose my job but on the other hand 'NW' was sort
of a mate and we had some good times on the door together, but for
some reason he was doing drugs and that was it, as far as I was
concerned.

That night in the club, I had to personally deal with every situation
that arose and it went off quite a few times that night and each time
'NW' wasn't there. One major event that happened did involve 'NW'
This Spanish guy was really pissed off at being charged a fiver to get
in after queuing for over an hour, he demanded to speak to the
manager. The manager came to the door with 'NW' and he told this
guy that the charge was a fiver and that's all there is to it. After a lot
of moaning the guy said to the manager

'Remember my face because I will remember yours'

This was a threat and when these guys make threats, it can be death
threats. For some reason that I still do not know to this day, 'NW'
went out of the club and came back a few minutes later with this guy
and gave his fiver to the manager. I thought what the fuck is he doing;
this guy just made a serious threat and 'NW' just brought him back
into the club?

The manager was not impressed by what 'NW' had done but let it go
because things were getting hectic on the door. The Spanish guy had
only spent two minutes in the club and came back up stairs to the
manager. He stared him in the face and said,

'Remember my face' and just walked out.

'NW' had really fucked up this time, supposing the Spaniard had
knifed the manager. How could he justify bringing this guy back into
the club? The drugs had changed his character totally and this was an
example of how dangerous things are going to be if I don't sort things
out.

'NW' was acting soft and I knew it wasn't because he hadn't any
bottle because I knew that he had been there many times before, but
the gear, whatever he had taken, had effected him and that wasn't
good in our line of work. We weren't there to socialise, we weren't
there to chat up girls, we were there to work and that was all I wanted
to do and 'NW' was acting unprofessionally. When the end of the

night came, we left the club to go get our wages. As we came away, walking down the street, 'NW' pulled out another joint and was smoking it on the way back to the club.

I turned round to him and said,

'What are you playing at? Smoking dope. What is the matter with you?'

I didn't even want to be walking down the street with him like that; I didn't want to be pulled up by the old bill. How would that have looked? Two doormen walking down the road and one was smoking dope. What would they have thought about me? That I was taking it as well? I was, and still am, a responsible person with a clean-cut image. I didn't need that.

He carried on smoking. He wasn't really interested in what I was saying. He said,

'We had a good night, didn't we?' and shook my hand, which we always did at the end of a session to sort of like, thank each other for watching each others backs, but I still wasn't happy!

When we got back to the club the club where the boss was working the boss took one look at 'NW' and he knew. He knew the score. He just needed to look at him. Maybe he had a whisper from one of the managers, I don't know, but you could see it in 'NW's eyes.

The boss was due to go abroad in just a few hours so I just said my goodbyes and went home. Anyway that was the end of that until the following week. Its really hard and a real down side of working on the door, when you're working with somebody who you think you know and who you think you can trust, but something comes along and lets you down. You just can't afford to be let down.

Anyway, after the death threat incident, I really knew that 'NW' was letting the side down, it was becoming dangerous. I had to back him up, being a fellow doorman, but it wasn't a situation I was happy with. So a few days went on and I knew that I was going to have to deal with 'NW' It was coming to loggerheads and we were going to have to sort it. It couldn't go on like that and I couldn't work with him anymore. Now, I didn't have 'NW''s home telephone number, he said he wasn't on the phone, he just had a mobile phone and I didn't have that number, he kept it pretty quiet and I didn't have his address either so I had no way of getting in contact with him.

So I sent a message through the ranks, through one of the other guys on the team, to tell 'NW' I wanted a word with him, and that I had a problem with him that I needed to speak to him about. The guy I spoke to said that I was best going through the head-doorman, let him know first what the situation was, because they would rather him deal with it than let me and 'NW' get together, he didn't want us coming to blows. So I did.

I spoke to the head-doorman and he told me to forget about it that he would deal with 'NW' himself. I told him that I still wanted 'NW' to contact me. I didn't want to turn up for work on Friday night with it all hanging in the air, we needed to get together and sort things out. I knew 'NW' wasn't going to be too happy about it, obviously I was threatening his job, he did three nights a week down there and it was pretty cushy for him, but that wasn't my problem. I didn't want to see him lose his job but I didn't want to work in a dangerous situation either. It needed sorting.

Now 'NW' had been working at the Aslas club about four or five weeks before I started, so he had got his foot well in the door. It may have been the case of last in first out so I had to tread very carefully; I didn't know what the set up was. The other thing I was concerned about was that I didn't know if the whole security firm were on drugs. Maybe I was the only clean one there.

I got a message that 'NW' wanted to speak to me through the other fellow so I didn't really know what it was all about. I expected it to come to blows and I had a feeling it was going to be a team effort where they were all going to stick together and that we were going to kick off with me. I didn't really give a shit anyway because it had to come to it and I couldn't keep working in that environment.

I made it quite clear that I wanted 'NW' to come round my house so I set a time. There was no point in me going round to his house because his wife and kid would be there and if it went off it wouldn't be nice for them. That's not my style. I wanted him to come round my house where nobody had to put on a show for lads on the firm, his family or the general public. I made a time for him to come round when it was just us there. There was just me and my eldest son living there and he would be out.

'NW' phoned me before the day and all he said was that he wanted to see me. I told him to be at my place at six and then he put the phone down, there were no pleasantries. I knew at this time that my boy would have gone to his youth club so he was out of the way and it was just 'NW' and me.

Now I didn't know what to expect or how 'NW' felt about things. All I knew was how I felt. I knew the other lads were probably winding things up a bit so that we would come to blows but I also knew that 'NW' would not want to fight me, up until now we had always been friends, but I still didn't know what was going to happen. A few weeks before he had gassed somebody and the last thing I wanted was to get gassed, I didn't want CS gas in my eyes. The thing that made it worse was that I had just had quite a heavy operation on my groin the day before. (*no it wasn't a penis enlargement - although I do need one, it was a vasectomy!*) I was in extreme pain and could just about stand up straight but I was ready to go for it all the same, I had no choice, I think I did expect 'NW' to gas me.

'NW' turned up and I asked him whether he wanted to go for a drink to the pub round the corner or whether he wanted to come in. He told me because he didn't drink that he might as well come in, but of course this made me weary straight away. I assumed that he didn't want any witnesses around which made me think he was going to gas me.

He wouldn't have done that in a local pub with so many witnesses around but inside closed doors it could happen. So I thought right here we go!

Now prior to this, don't forget, I had it in my mind that 'NW' may have some gas on him, I had brought some very high strength bleach which I had poured into a cup and had on a table next to me when we sat down. If he did reach for his gas I would have thrown the bleach in his face then grabbed him and bitted a piece out of his face.

As it was 'NW' came in and we had a long chat. I tried to keep on top of the situation. I told him I wasn't interested in what any of the guys had said to him, all the shit-stirring and back-stabbing, I was just telling him as it was, that I felt he was a liability and he wasn't pulling his weight under the influence of dope.

'NW' explained his situation to me. He was having problems in his personal life, the details of which I need not write in this book.

I then understood why he was doing drugs but to me it still didn't make it right. 'NW' apologised for all that had happened and we shook hands. Even whilst shaking hands I kept my guard up ready to go for the bleach. The thing is I believe that when you give somebody your hand to shake, it is the only time in life, except for lovemaking, training or dancing, that you voluntary give another person part of your body to hold. This makes you vulnerable and in most cases you shake with the hand, which is your strongest, your fighting hand. So even as I shook his hand I was ready to go for it in case it was a ploy to get me off guard.

Anyway we shook hands and it was dealt with. 'NW' gave me his word that he wouldn't smoke dope, or do any drugs at all prior to work or at work. He did ask me if I minded him doing drugs after work which I said was up to him as long as I wasn't walking down the street with him at the time for fear off getting pulled up by the police. I didn't want to lose my Door Supervisors license also I didn't want to be associated with drugs of any kind.

It's very ugly when this sort of thing comes between two doormen because with doorman if it does come to blows, it can get very, very heavy. I had just wanted to deal with the situation the best way possible, avoiding any heavy stuff, then continue working and to make sure my life was protected as well as 'NW''s.

We worked together the following weekend and everything was fine.

I did find out that prior to 'NW' coming to have a chat with me, he had worked the Tuesday before and had let the guy who had threatened the life of the manager back into the club again. 'NW' realised after the chat what a mistake he had made and it was down to the fact that he was smoking dope.

He did say in all honesty that night that he didn't take anything else; that the high he was on was down to the dope, that it affected him differently to other people. Although I've never heard of that with anyone else before I did believe him. And he gave his word that he wouldn't touch the stuff whilst working again and that was good enough for me. When I give my word on something –that's my bond and I stick to it, so naturally I think other people are the same as me but every now and then I get proved wrong. A few nights after that he didn't show up for work at all. It was a busy night and I was left to supervise the club on my own. His word wasn't worth shit! The same thing goes for most other people with a drug problem.

You should never be under the influence of drink or controlled drugs whilst working and employed to prevent crime. Not many doormen see themselves as being in the business of crime prevention, but essentially that's what it is.

Many doormen I have met or worked with who are doing drink or drugs are beginning to lose the plot and need to get a grip on themselves.

Some of them are doing it because they are losing their bottle and need a boost to get through the night and others just can't cut the mustard with out drugs. Losing your bottle is not a crime, I've lost mine on a few occasions, just accept it and get out before you or some other person unnecessarily gets hurt, but drugs are a crime.

I once took a kicking from 10 blokes on a dance-floor and ended up in hospital coughing up blood and having a tooth extracted from my gums. While this was happening, my fellow doorman was discussing the finer points of cannabis with a customer and claims to have not seen me go to ground. Maybe he thought a Rugby scrum was a natural occurrence in he middle of a dance-floor. He was stoned and suffered dearly for his temporary attack of blindness letting me down.

It's funny how the club environment changes the character of some doormen. Its like they are a boxer getting into the ring to do what they do to please an audience but the difference with being a doorman is that all your actions have an affect on others in some way. The problem is that some doormen carry on their role once work is finished and another problem is that others want to carry it on after your work has finished.

Whilst working within the club environment, you have the security of the walls around you, just as you have the shell of the car protecting you when driving if road rage occurs but things change when its time to go home. The doors close around you, and you're alone. As you read earlier with 'CC' who found this out the hard way.

I know one doorman who acts the big fish in his little pool whilst working on the door, protected by the club environment, yet when driving home alone; he will not even glance out the side window in case he gets a confrontational stare. He really abused his position of doorman whilst protected by the other doormen, the Law and the comfort zone of the walls around him but would shit it when he stepped outside of that environment.

There have been many a time where customers have been abused by doormen so the customer then lays in wait for the doorman to leave the club and make his way home to get his revenge. If you are a good guy, its unlikely that you will ever find yourself in that situation, however even as a good guy, you may still find that you are followed for a little more than your autograph.

I was working for a very well known club in London's Soho, just to give you an idea as to its size; it was a Plc owned club.

I spent a few years there attempting to keeps the low life of Soho outside of its doors.

The club preferred to operate a low profile security policy where customers were not allowed to be searched, in order to give the entrance of the club a trouble free image, in turn creating more business.

The reality of the situation was that it became an ideal 'Safe house' from which drug dealers could freely walk into and operate without being stopped and searched at the door.

Morally, I could not accept the fact that I, as security, was letting controlled drugs be sold on premises that I worked.

I'm no saviour for all mankind but having been exposed to my sister's problems all my life, I felt that I should contribute something to ridding society of this shit.

I soon figured out who the bad guys were and refused them entry to the club, in a polite and courteous manner, explaining my reasons for doing so quite clearly. This guaranteed me becoming very unpopular. I don't like being unpopular but when it comes to drugs I don't mind making an exception.

Anyway, with their operation now redundant - the offer of bribes, to turn a blind eye, began.

When they found that their offer of £500 a week for me to turn a blind eye, to their drug dealing, was falling on deaf ears, things began to turn nasty. The threats of beatings and knives being flashed before my eyes were just the start. Just think about it? £500 per week to do nothing! What would you have done?

I feel so proud of myself for making that choice and hoped that this made a difference to some young kid's life. It won't stop the drug dealing because the same shit will happen–just in a different place.

The drug dealers and their little workers who would cut your throat for a bit of gear decided to make my life hell.

It progressed to them waiting outside for me, knowing that I had to leave the building eventually, so I played careful and left via the fire exit. They must have used every intimidating ploy in the book in order to scare me off the door of that club.

Just like an old gangster movie, I even had a white Mercedes packed tight with big guys following me home to intimidate me even further. So a few detours, phone calls, and a waiting party to welcome them at the other end soon solved that one. One of the wankers thought himself a bit tough and threatened to come back on my family so I opened up his throat with the sharp edge of flattened tin can and told him that next time it will be his eyes. The handy thing about studying the fighting arts is I knew not to cut the Carotid artery or jugular veins and just how far to go. I don't choose to get in a fight with anybody so if I am then it's because someone has chosen to have a go at me! They will have then forced me into violence against my will and will have to pay a high price for that. I have not previously been able to declare this incident through fear of upsetting my mother who was peace loving and frightened by me being at risk through violence. Now she has sadly passed away I will bring out in the open some more of the events that I previously would not speak of due to my respect for my mother. There is not enough room in this book to tell all so I will include them as I am able in future publications, as long as I am not going to get myself locked up.

Anyway getting back to the pricks that followed me that night.

The strangest thing about it all was they seem to know my every move and I thought it might only be a matter of time before I get the homely visit if I didn't sort matters out there and then, so I did.

It turned out that the other doorman, who I was working with at the time, was also working for them, without my knowledge. He liked both drugs and money so their was no way he could refuse an offer of £500 and all the drugs he could absorb so he went to work with them without me knowing.

Once trust goes out the window, its time to Part Company, so I left, but I can live with myself knowing that I was the good guy and left when I decided to, rather than someone else deciding for me!

Obviously I did not remain friends with the 'Judas' of a doorman that I worked with then. In fact the situation between us got so bad that I went to an arranged meet with him expecting a big time kick off with him and his mob. I went fully tooled up and alone because it was something that had to come to an end one way or the other. I had no choice really because other than that I would have been sitting indoors, never knowing when the petrol bomb was going to come through my window or the front door being kicked in, so I went to deal with it face on.

As it turned out there was no mob to greet me, just the doorman on his own who wanted to call a truce. He came back to my house to discus it all and I unloaded my toolkit and we agreed that we would just stay out of each other's lives. I went to my shed in the back garden to put my tools away and as I came back out he was standing there with a 357 Smith & Wesson pointing in my face. He fired the gun and the bullet sunk itself into the grass just beside me.

To this day I do not know if he was a bad shot and tried to shoot me or he intended to miss. He said that he wanted to show me that he had the opportunity to kill me if he wanted and fired the gun to prove that he didn't want to harm me and wanted a truce. I can think of much better ways to prove a point than that, but that's what happened. Another reason that we agreed on a truce was to save any unnecessary bloodshed. We both had children, friends and family that could easily have got caught up in the crossfire and that's something that neither of us wanted.

As for the gun, and how I knew it was a Smith & Wesson! It was mine. I owned it quite legally and had it out because I was getting it all ready for its disposal by the Home Office. I had been a licensed firearms owner for 12 years. I applied for a licence around 1986 so that I could have a real gun indoors and use it to shoot paper targets at the firing range (yeah right!). I had a gun for the same reason that most of the other people I knew who had guns; it was another form of security.

Not many licensed firearm holders had guns to get off on shooting paper targets, there are some but you will find that most of them just wanted to have a gun indoors. I was not bothered at all when the guns were finally ordered to be destroyed. If it saves just one person's life then it is justified but while they were available, I wanted one. I must say that I am of the opinion that taking guns away from licensed firearm holders will not reduce criminal activities with guns in any way. There were plenty of us licensed guys queuing up to hand our guns in but I didn't see many villains handing in theirs.

Another thing about having a gun indoors was that whenever I was out of a night working the doors and my mother was home alone or looking after my son Jamie, I always made sure that she had access to the gun if she ever needed to protect herself. It's very easy in a perfect world to criticise me and my actions but I live in the real world where people break into houses, they do rape women on their own and they do kill children. A bullet would put a stop to any of this. I have since replaced my gun with a dog, my lovely Staff Kimba. Although dogs don't know any Kung fu, they can rip an intruder to shreds if the situation arose.

Obviously it would be much easier to stay home and protect my family myself but the bills still had to be paid so I had to carry on working.

I was so sick of it all, I wanted to get away from doorwork so badly but I needed the money. Every doorman has a shelf life and once you pass your sell by date you have got to think about getting out of the game.

I had always dreamed about being a writer and would spend hours on my computer typing out my war stories. I was living on my own and would come home night after night and have nothing but four walls for company. Living alone can be very depressing if you don't make good use of the time, however for me it was an ideal setting to write in. The time slot from 3am to 7am allowed me to write without any interruptions whatsoever. I loved it.

I always dreamed of writing a book and it rewarding me by getting

159

me away from doorwork, which it ultimately did, but not in the way I thought it would. I will enlarge on that later.

I realised that I was getting older, slower, fatter, weaker and was also losing that sharp edge that always made me feel strong when situations occurred. I also think that as my children were getting older, they began to realise what I did for a living and that kind of embarrassed me. What sort of example was I to my children? I was a fucking sad doorman. I don't mean this to be detrimental to other doormen who choose and enjoy this occupation, I'm referring to me and what I wanted from life. I didn't want that for myself or that be the role model that my children could look up to me as being.

I remember once one of my kids was asked by their teacher what their dad worked as? My son said,

'He's a bumper, miss. When someone starts a fight in a pub my dad beats them up' That's how my children saw me.

The teacher replied,

'I think you mean Bouncer! Not bumper'

I decided that I was going to save up enough money to take my four kids away for the holiday of a lifetime so that they would always have something special to talk about when asked about their dad, rather than saying, 'He's a Bouncer.'

I took them away to Disneyworld in Florida and gave them the best time of their lives. I was temporarily stuck in the role of a doorman so decided to make it work to my advantage. Every night, every fight, every lucky escape of near visit to deaths door was made just that little bit more worthwhile just to take my kids away. After Florida I would never return to the door again. But as Sean Connery said, 'Never say never.'

I was back on the door within two weeks of returning back home.

I just couldn't handle the lonely company of my four walls night after night so back to work I went.

I went to work in a different club, which was pretty calm compared to what I had been dealing with for the last year, so allowed me to chill out a little bit.

I decided to put aside my project of war stories on the door because it just did not feel comfortable whilst I was still working within that environment and decided to do a book on female self-defence. I was teaching female self-protection to a few girls' schools' and thought this would be a good subject to cover. It took me two years of research and thousands of hours at the keyboard, within my unwinding period after work each night, but I finally got there. My first book was called *'Dogs don't know kung fu'* which I dedicated to my mother, sister and all others who have suffered in silence at the hands of a male. I expected this to be the only book that I ever released. That was it. The book I had always wanted to do was now out of my system and I could now go on and do something else with my life, or so I thought. This is now my sixth book and I have another dozen already lined up. It's funny how some things work out?

Anyway, to get back to the point. I was still working on the door.

I truly wanted to deal with conflicts in a non-violent manner so I could feel that I was doing my job properly, to the best of my ability and in many situations it would have been a big mistake on my part to engage in a battle. Some situations are like bombs waiting to go off. You have to defuse them to save a lot of damage from being done and it also prevents a chain of events escalating into something uglier. You are supposed to be preventing people from committing a crime, without becoming Judge and Jury. If you work with one or more security, you should be able to put this into operation quite professionally, without you appearing to take either parties side. Don't make the mistake though of defusing one situation, only to find that you are in a worse one.

'CF' was a racist doorman who disliked me. My crime was being white. To this day I don't believe that 'CF' realised that he was being racist. Many black guys don't. Anyway the fact of the matter was that he tried to make my life uncomfortable. Not openly, but in indiscreet little ways. Silly things like not swapping with me on the door when I needed to go and piss. Slipping off before the final few customers were out the building and stuff like that.

If you have ever worked the doors you will know that clear out time is one of the times you really do need help. There was nothing racist in these actions alone, it was other things that brought it to light, like when I broke up a fight once. It just happened to be two black guys fighting each other over a girl. It was no big deal and they calmed down after I spoke to them. Straight after 'CF' came over to me and said,

'In future, stick to dealing with your own kind' I replied,

'What do you mean? Fat blokes'

I knew exactly what he meant; I just wanted to hear him say it.

He sucked air through his teeth, which is a kind of disrespectful 'tut' and said,

'Leave the black guys to me, you don't understand our ways'

'Don't be so fucking stupid', I replied,

'I just see two guys fighting and I break it up, I don't see black, white, Asian. I just see a problem that needs to be dealt with'

He walked off.

The rest of the 'Brothers' that formed our team were fine. It was just this prick. With a minor problem like this I let things go for a while just in case I'm being paranoid. I also like to watch out for repeated patterns. If something keeps repeating itself then it's a bit more than coincidence. 'CF' was the most switched off doorman in the world. He would actually get on the dance floor with the customers and dance to his favourite songs. Have you ever seen anything like that at a club? A fight would happen right next to him and he wouldn't even see it and certainly wouldn't stop dancing to break up a fight. He was a real arsehole. When you were working with just him as the only other doorman, you were working alone. Working on your own is not nice, its when necessity and stupidity have become one, but I have had to do it many times.

Later on that evening the two guys pushed a few more beers down their necks and began to squabble again. 'CF' was on the other side of the dance floor, with his lights switched off, so I attracted his attention by switching on my mini maglite and flashing it across his eyes. He looked over at me and I pointed to the argument that was happening right beside him with the two guys.

He walked over to them and tried to lay on the *'Yo Bro, what's happening'* but got told to fuck off. I began to make my way over to them and saw one of them pull something from his back pocket. As I got closer I saw that the guy had a knife, which was not visible to 'CF'. I managed to grab the guy and pull him in a backward direction through the fire doors. I detected that by now 'CF' was behind me and had realised that I had just saved him from getting stabbed. I threw the guy over onto the pavement then heard the fire doors slam shut behind me. 'CF' had shut me outside with the knife guy and had positioned himself safely within the secure walls of the club. It would be nice here to tell you about the fantastic knife fight that I had with the guy outside but nothing happened. He just got up, picked up his knife and walked off. I then had to walk all the way around the building to get back into the club.

I approached 'CF' and said,

'You fucking c**t, what do you think your playing at shutting me out of the fire doors when I'm faced with a guy holding a knife, after I just saved your fucking arse'

He was shaking like a leaf and quivering as he replied,

'I'm really sorry man; I just panicked and didn't realise what I had done. The knife thing scared me; I just didn't know what to do. I'm sorry man, please let me make it up to you' I replied,

'Just stay out of my face, you're a wanker' then walked off.

The other guys had seen what happened and had words with 'CF'.

I don't know exactly what was said but he came over to me again and apologised asking if we could start afresh, putting this incident behind us. I said,

'I can understand you feeling scared, it happens to all of us sometime, but don't ever do anything like that to me again. You can stop all this racist bollocks as well, I don't like it'

We shook hands on it and I let it go but I never trusted him again. Working with him was fine after that. I think he realised that his dislike for me was an unjustified chip he had on his shoulder.

When you condense a few stories about conflicts on the door it sounds like it was fight after fight, night after night, but it wasn't always like that. We would take Christmas day off!

There were plenty of funny things that happened like when the night Bob Geldof came in the club and Micky the doorman asked him how Paula Yates was. Micky didn't know that Bob and Paula had just split up over her involvement with the late Michael Hutchence. Geldof was well pissed off. Paula has since passed away as well.

There was also the time that the band 'Beautiful South' asked me to take them to some safe clubs in town. They were recording in London that day and wanted to dance the night away without any problems from Londoners.

I took them to the Hippodrome where I knew Pete, who was in charge of the door at the time. He made us all welcome and the band were having a great time chilling out on the dance-floor while I discussed a long term security plan with their manager. Next thing I know a couple of Algerians that I had kicked off with at the 'Aslas club' spotted me. Let's face it, I'm not too hard to recognise. In no time at all the two Algerians expanded tenfold and they steamed me. If it wasn't for the high quality and promptness of the Hippodrome team that day I would have been a gonner. It was just one mass kick-off. We managed to get the band out the back doors safely but they were not impressed. Needless to say I never worked for them again.

It didn't affect my career in anyway, the week after I was doing personal Security for Bruce Willis. He was doing a secret gig in London with his band and I was asked to do his VIP security.

It was my job to seek out any crazy stalker types that were likely to create a problem. Around that time period I was getting booked to do plenty of VIP security and had some brilliant offers come my way but that would have meant not seeing my children grow up. I was only getting access to them eight days out of each month as it was so could not risk losing that just for high profile work. Your kids are only kids once. Security will always be there.

If you get the right kind of security work it can be a great lifestyle and you can get to meet and work with people that you would never normally cross paths with.

I was doing personal security for Gary and Martin Kemp one night and Gary was doing some songs to promote his solo career. Martin stars in 'Eastenders' now as Steve.

I had been a fan of Gary Kemps songwriting since his days with Spandau Ballet. Both him and Martin are better known these days for their portrayal of gangsters Ronnie and Reggie Kray in the movie 'The Krays.' Anyway the strange thing was that the guy who was sorting out his guitars and bits for his performance was Scottish so I started talking to him about my days as a child in Scotland. We realised that we lived down the same street and had met as kids around 28 years ago. It was really spooky.

We were both friends with the only Chinese family in town.

Who would ever think that we would ever meet up under those circumstances, all those years later?

I had some nice pictures done with Martin and Gary that night, for inclusion in this book, but afterwards Micky put the camera in a McDonalds bag in the back of his car for safe keeping. A week later he cleared his car out and threw the bag away, forgetting the camera was inside. It was never recovered.

A few days after doing the Kemp brothers I was working again looking after Noel Gallagher one night, then Bryan Adams the next, so as I said it can be a wonderful lifestyle if that's what you want to do but you have to be good at your job and sacrifice family and friends to do it. I wasn't prepared to do that long term. Noel Gallagher, Bruce Willis, and Bryan Adams couldn't even tell you

Watching Noel's back – just another nights work

my name now but the people closest to me could. Family are the ones that count. Sometimes an incident will happen that sticks with you because it was strangely funny like the night I got beat up by a girl! How many big tough doormen would admit to that one? This is how it happened.

The club was full and nobody else was allowed in. It was near to closing time and we just wanted to get finished up and go home.

A rather drunken girl came up to the door and demanded entry. She was a very big-built, stocky, lesbian, looking for her girlfriend who she had argued with in a neighbouring club only minutes earlier. For some reason she thought that her girlfriend had entered our club but I had been on the door for the previous hour and knew that nobody had come in. I explained to her that it was not possible that her partner was in our club and suggested that she went back to the club she had come from. Next thing I know she had launched a blinding uppercut to my solar plexus knocking me through the doors. The Walkie talkie went flying out of my hand and I went crashing backwards into the wall. I was more shocked than hurt. Admittedly I was not prepared for this one and got caught out, but the surprise of a female throwing this type of punch was unexpected. I started laughing at what I thought was a funny situation. This triggered the girl into an aggressive rage and she ran in the doors, grabbed a glass and tried to ram it in my face.

This I took more seriously. A prompt kung fu style double deflection removed the glass from her hands then I used a bear hug style hold to smother her continuous attempt to punch fuck out of me. If it was a guy I would have broke both his hands for attempting to glass me but I just could not do this to a female, even one that fought like a man. I dragged her out the door, which was a feat in itself being as she matched me pound for pound in body size but eventually I got her out and bolted the door.

The manager and some of the other staff had watched the whole thing on the CCTV and were taking the right piss out of me. They were giving a running commentary over the Walkie talkie, which I didn't hear at the time because I was struggling with this girl trying to save my bacon.

One thing about door-work is that it is not compatible with a stable relationship. Door-work and stable relationships just don't last. Mitch was a fellow doorman that I worked with who just could not get himself a regular girlfriend. I don't know whether it was down to ugliness, shyness, or he was just to tight to take someone out, but he couldn't get a girl to stay with him. He did do pretty well though at pulling lots of different girls but they would never seem to stay around for long.

He later confessed to me that his game was advertising in the local lonely-hearts column for partners. They would see his advert that read something like,

'Good looking, lonely guy with disposable income wishes to meet girl for long term relationship etc'

All Mitch ever wanted was to have a permanent girlfriend. His world fell apart after his wife had an affair with one of his mates and set up home with him. Mitch was a faithful sort of guy who never strayed from the path and just wanted that comfort zone that he enjoyed so much, back again. Every weekend without fail he would come into work with a list of 15-20 women that had left messages on his answerphone who were interested in going out with him.

Some were lone parents, others were single or divorced and a few were dirty slag's who were cheating on their husbands but to Mitch, they were all worth a shag.

He would phone them up and arrange for them to come down to the club on different nights so he could check them out without him even leaving work. They would come into the club and spend the night getting pissed and at the end of the evening Mitch would kindly drop them home. The rest you can guess.

If he could find one that liked being filmed he would get out his camcorder and film them doing themselves with all sorts of toys and vegetables, then bring the camcorder into work the next night and share his experience with anyone that wanted to watch it. He was a nutter! Sometimes he would play his videotapes on the night-club TV screen after the club had closed and all the staff would sit around drinking and making fun of the tarts that Mitch caught on camera. The majority of the women on film came from Essex; yeah you've guessed it Essex girls! He ended up with videotape with about 30 different girls on it ranging from about 30 years old to around 60. Each and every one of them knew they were being filmed by a guy that they had only just met through the lonely-hearts column and were more than happy to be filmed. I don't think however that they expected to be featured on a nightclub screen doing their stuff. Things backfired for Mitch in the end because he actually grew to like one of the girls but just couldn't carry on with her after letting all his mates watch his video of her shitting in his frying pan. Luckily it was non-stick!

Another funny thing that he used to do was tell the girls that he would pop round to see them after work if they described to him over the phone what pleasures were in store for him when he arrived. He would phone them just after 7pm when the calls were free on his mobile and the girl at the other end knew that she had about 30 minutes to tease and tempt him before he arrived at work. What they didn't know was that he had a hands free phone and also picked us as passengers on the way.

So there would be four of us in the car being forced to listen to his game of phone sex with this woman. We didn't have a choice as to whether he played his radio or made phone calls, it was his car. He was a real strange one!

Egg on Your Face

Most of my life has been associated with thugs, and violence of some shape or form but this book would not be complete without talking about some of the mugs I have met. For those of you not familiar with the term mugs, it simply means idiots! People who get egg on their faces. In the martial arts world we have a saying that fits nicely along side *'Egg on your face'* it goes *'Empty heads have long tongues'* and believe me, those with empty heads talk bollocks and get themselves into a lot of trouble doing so!

I was working at one club in Southend where a lot of younger doormen were employed. I suppose on average they were 10 years younger than I was and to be honest, I had forgotten more about door-work than they ever knew. I was seen as the mature one out of the pack and was often offered the position of head doorman but that's something I've never been into.

One of the lads 'IT' was your average cocky jack the lad who was badly suffering from insecurity within the job. He always felt that he had to prove himself by mouthing off at customers and saying what he was going to do to them if they didn't act sensibly. The problem was that he only ever exercised his position if it was a girl or guy that was smaller than he was.

'IT' did not get his nickname from 'Information technology'; he got it because he carried out so many 'Idle threats'. The thing about 'IT'. was that he was a really nice guy deep down and would do anything he could to help you out but he was insecure about his ability to deal with confrontation. As a result he ended up doing Speed to give him the balls to get through the night. This later progressed to doing Coke.

The knock on affect of doing Speed and Coke is that you begin to suffer badly from paranoia and boy did he suffer. He gradually went through each of the doormen one by one challenging them in different ways to prove he was tougher than they were. He had been studying kickboxing for a while and would demonstrate it to the other doormen and even customers at every opportunity he could. He would really show himself up but obviously he couldn't see that himself, he was a mug.

It was my turn to become the target of his challenge because I was the only one left that had not backed down to him. Every single one of the other doormen had backed down to him and in turn would practically bow to his command. He was chasing the title of head doorman and the problem with this was that he saw me as being in the way of that. It was not the case. I just wanted to do my nights work and go home and was not interested in any titles or stripes on my arms. I got a whisper from Joey, another doorman, that 'IT'. was going to sort me out because I looked at him the wrong way a few nights previously. Its not something that I knowingly did but to be honest I didn't give a toss as to what 'IT' thought.

'IT' had supposedly said to Joey,

'I'm going kickboxing on Wednesday night and after that I'm going down to the club to have a run in with Jamie and chin the fat bastard for screwing me out'

I knew what 'IT' was doing; he was hoping that I would get the message and be too scared to turn up at work on Wednesday and in turn he would come out the hero! But sadly I don't take too kindly to threats and off to work I went.

As the night progressed all the other doormen were dying to see it kick off and had wagers on what the outcome would be. After a few hours of listening to the schoolboy comments like,

'Are you worried' and *'I see your still alive then'* I had decided to act. I phoned 'IT' on his mobile and said,

'Its Jamie, I hear you want to have words with me', he replied,

'I'm at the gym training for my title fight, afterwards me and a few of the boys are gonna go down the club. I could have sorted this all out face to face if you were there'

He threw in the 'title fight' bit hoping that I would see him as someone out of my league. Bullshit like that doesn't impress or scare me one little bit. I said,

'I am at the club, I'm working, and I will be here all night waiting for you and anybody that you bring with you. I don't need any back up, I will be here on my own'

He went silent for a second. He obviously didn't realise that I had gone to work and must have thought I was phoning from home.

He screamed back down the phone,

'You don't fucking scare me, you have called it on and I will be down to sort it out' I ended the conversation with,
'Yeah, whatever!'
I thought fuck it, I don't really have any other choice other than to deal with it head on. Backing down was not an option for me. There are times when backing off is the right thing to do because you are in a no win situation but that was not the case here. A one on one situation was the least of my worries here. The worst-case scenario that I had here was that a few of them would steam me but if that was to be the case, they could do that anywhere and I would eventually get them all back one by one, so I stayed on the door to face the music.

I would have hammered him badly in a one on one and he knew it. So I expected an ambush or to get squirted as his way of going through with it after telling the other doormen and the guys down the gym what he was going to do to me.

I waited all night and nothing happened. I went home and found a message on my answer-phone from 'IT'.

'Jamie, mate, listen about all that business on the phone tonight. I've got loads of problems at home with the wife and stuff and you just caught me at a bad time. Sorry I blew my top; I was out of order. Let's get together and have a pint so I can apologise'
I thought, what a wanker, he's brought all of this on himself for no reason at all except for his own ego. I just laughed and went to bed and gave him credit for backing out. It takes a lot of guts to do that when you have called it on in front of all our mates. The next night he came to work looking like shit. He walked straight up to me pushing his handshake out as far forward as he could. I just smiled and shook his trembling hand. As he started apologising again I said, *'Just forget about it, it's not a problem. You look like shit what's a matter with you?'* he said, *'I've been up all night expecting you to come steaming through my door. I know that's the sort of thing you do. I just wanted to catch you before you came up my path so I could apologise before to kicked my door in. My wife and kids were indoors; you know what I mean. I thought after I didn't hear from you after I left that message for you at home, you might still be pissed off about it and pay me a visit'*

171

I replied,

'Don't be so fucking stupid. I don't do that sort of shit anymore also never would if family were involved, that's not the done thing; I only do it if I have a major problem to deal with which I see as a serious threat. Our situation was just something stupid created in your own head and blown all out of proportion by you. Let's just forget it,'

'IT' went away with his tail between his legs feeling a right idiot and so he should. I'm an easygoing guy who can take situations like this and forget them the next day. 'IT' will remember though for the rest of his life. I hope for his sake he doesn't try it on next time with someone less tolerable than myself. Every time I saw 'IT' after that incident he couldn't do enough for me.

Ponytail was another idiot that I worked with. I actually liked him as a doorman because he was well mannered and polite to customers and did not look like he had just been released from prison. Prior to working with Ponytail I had 15 years on the door and 20 years within martial arts but I was fairly new to this establishment so nobody knew anything about me. To the other doormen I was just a fat bloke who looked hard but was probably a bit slow when it kicked off. I never got into conversation with Ponytail because he was a martial artist who would jump up and kick the lights hanging from the ceiling and other martial art stunts to try impress the other doormen and staff. I'm used to being impressed by the top guys in the martial arts premier league and Ponytail's efforts were pathetic. Don't get me wrong, his martial arts may have been the bollocks, I don't know, but the way he conducted himself was a joke. Ponytail also had **'Wannabe head doorman'** syndrome and tried it on with all the other guys as and when he could. One day I went to the toilet to drain the lizard and was greeted by Ponytail on my exit.

'Next time you want to go to the toilet you ask me, I'm the man here,'

He said as he stood there with his chest pushed out. I replied

'Fuck you Barbie! I don't take any shit from you so fuck off and play head doorman with someone else,'

He stormed off with his bottom lip sticking out like he was gonna cry.

Oh dear! I've made another enemy. Its something I was getting pretty good at.

Ponytail just ignored me for the rest of the week like a sulking child but always made sure I was within eye view when he did his martial arts stretching or imitation Bruce Lee strikes to the air. He was pathetic. I was actually looking forward to him confronting me with his martial arts bullshit so I could bite his fucking face off but it never happened.

The weekend came and a situation went down where I ended up in a scuffle with three blokes and my entire backup had disappeared on me. Oh how unusual (not). Anyway I handled it the best I could by sweeping the legs away from the biggest threat then banging one of the others on the chin putting him down. The third guy just stood there ranting and raving at me offering me out. Ponytail then arrived on the scene and totally mishandled the situation with this third guy. As he positioned himself into his martial art stance the guy picked up a pint of beer and poured it all over Ponytails head. Ponytail just stood there gutted. The three guys then just walked away calling out the usual *'Wankers, we will be back'*

When we walked back into the club my knuckle had swollen up from where I had hit the guy. It's not something that normally happens to me but it happened that day. Ponytail was soaking wet with beer dripping off the end of his nose. I said,

'I've done my fucking hand I am going to get some ice to take the swelling down' Ponytail said,

'If you come to train in martial arts with me I will teach you to punch properly so that you don't get injured like that. Did you see how that guy threw drink at me and I didn't move from my stance'

He had to try and justify why he looked like a raincoat. He still didn't have a clue that I had studied the arts extensively. He was using the arts as a way to try and be my mate because I just watched him getting mugged off.

I said,

'Listen, I know you have been telling people that I look too fat and slow to deal with a confrontation on the door but I just put away two guys in two seconds and would have done the same to the third one if he stepped within my space. You squared up to one

guy and he soaked you. What the fuck are you gonna show me?'
He walked off greeted by the other doormen and customers laughing at him. He never returned to work again.

Stinky was another doorman I worked with who looked as hard as nails and hardly ever had to deal with a confrontation because of his appearance. He was built like a brick shithouse and looked like a bulldog.

He acquired the name of stinky because he was a little soap shy. He stunk to high heaven. Over a period of time stinky built up a false sense of security because rowdy customers were never confronting him. They always preferred to kick off with the other doormen who looked an easier target. As a result Stinky actually had a false feeling of toughness. He gradually began to abuse this position until one day it came on top for him. He had found some guy snorting Coke in the corner of the club and told him to get out the club. The bloke was leaving with no problem but Stinky pushed it a bit further because we were all to hand and he wanted to impress. The guy was only tiny but completely out of his head. As he passed us Stinky said to him,

'If you ever come back to this club again I will personally give you a fucking slap'
With that the guy turned around to him and said,

'Come on then you fucking wanker, come outside and slap me now, I will kick your fucking arse'
Stinky just froze, he didn't know what to do, his bottle went.

To make things worse for him there were 12 of us doormen present when it happened. He looked a real idiot. Not because he had refused the challenge, but because he was the one that offered the guy out in the first place. The guy just stood there continuously offering him out until Stinky disappeared back into the club. Stinky kept a low profile for a while after that. I did feel kind of sorry for him but it was a lesson that he had to learn.

I think the last time I had an 'Empty heads have long tongues' situation was when I worked with 'DD'.

'D' was new to door-work but tried to create this wonderful life that he had on the door in his past. He could never quite remember the names of the places he had worked or the names of the other

174

doormen and the funny thing about it was he did the same with the martial arts.

He made himself out to be the best thing since sliced bread. I liked 'D' because he was mature in many other ways and I did trust him by my side but it was the things he said behind my back that showed him in a different light. 'D' heard from another doorman that I was supposed to be a martial artist of some kind and asked me what it was I did. I told him that I was not loyal to any one art because I was into the Self-protection thing, which crossed many boundaries. 'D' told me that he was a kickboxing instructor but had an extensive background in karate in the very area that I lived. I have a habit of changing the subject or avoiding the subject of martial arts as much as I can because after 20 odd years in the game it just gets boring. Especially talking to people with closed ears, and most martial arts people that I know just want to talk about themselves.

As far as 'D' talking about himself and the techniques he personally taught, it was good and valid stuff which gave me a bit of confidence in him if it kicked off and we were together, but he made it his business to go to another doorman and say,

'That Jamie is a fucking liar, he doesn't do martial arts, I checked him out in the martial art circles that I mix in and nobody has ever heard of him'

Who the fuck did he think he was, checking me out to see if I existed in the martial arts. I'm certainly not famous but it's pretty hard to search through the martial arts and not come across my name somewhere, even if it just someone saying I'm a fat wanker or something like that. Anyway this doorman that 'D' said it to was actually reading a copy of my book *'Dogs don't know kung fu'* and had it in his car. He showed it to 'D' who displayed the emotion of being gutted. His reply was supposedly

'Anyone can write a book about self-defence but where are all the techniques? There aren't any kicks or punches in the book and nobody is wearing Karate suits' Oh! how naïve he was!

I went on holiday that week so never got to see 'D' again because on my return I went to work elsewhere. It's such a shame that some people are so insecure about themselves that they have to try and find fault in someone else. 'D' was a nice guy who unfortunately

175

didn't look hard enough to find out that I, and many others like me do exist within the martial arts world.

I think 'D's problem was jealousy. The old Green-eyed monster!

I'm sure to this day that 'D' is still spreading stories or denouncing my existence whenever my name crops up in his circles. How sad.

Thoughts

In my final two years on the door I think I went through my own private hell due to all the things that were going on in my private life. At one time I actually believed that I was going to die on the door, which is why I wanted to get my kids to Disneyworld before it was too late. The numbers game dictates that sooner or later a knife, bottle, or gun will win the day and I would be leaving the door with the sheet of death covering my body. I believed it so much that I asked my good friend Pete West to see to it that my first book 'Dogs' was released if anything was to happen to me. Pete was training with me privately week in week out and got to know me very well. He also got to know all the stress that I was going through and was a good friend that I could talk to. Pete was very instrumental in the actual printing of my first book. Without his help and backing the book would not have been put to print. So Pete if you are still reading my books, thank you for your unconditional help and friendship!

Another person, who I also owe a thank you to for the encouragement he gave me in releasing 'Dogs', was my pal Geoff Thompson. I sent Geoff a draft copy of 'Dogs' to ask his advice as to whether he thought it was worthy of being printed or not.

Rick Young, Geoff & me 1994

Geoff was knocked out by the book, he loved it and was probably the deciding factor as to whether I went ahead with it or not. If

177

Geoff had told me that it was crap I would have found the rejection at that time very hard to deal with and would have put the manuscript in the bin. His reaction to my manuscript was to have an effect on me that he could have never envisaged. He had a few books out himself and was becoming 'The man' for books on self-protection.

Many people in his position would have panicked if someone released a book that could be seen as possible competition or something that threatened to take a slice of their cake, but not Geoff. He gave me the boost of confidence that I needed to go full steam ahead and further reinforced his feelings by writing the books foreword, and once again, just as with my friend Pete West, Geoff's help and advice was unconditional with nothing sought in return.

My retirement from door-work was both hard and easy. It was hard because that was the life I knew and work was plentiful but I had to change the direction that my life was going in so I knew a choice had to be made. As I've said before permanent relationships and door-work just do not work so unless I wanted to be on my own for the rest of my life I would have to change my line of employment. Whilst I was on my own I dedicated the time to writing my books.

I left doorwork and began to promote my book. This was the only book I ever intended writing and had no plans to do any others, but 'never say never.' I sat in front of my computer again to write some more. I gathered together all my notes, diaries, and memories and put together my next book, called *'Old school – New school'*, which was a guide to Bouncers, Security and registered door supervisors. The foreword for Old School was written by the very well respected bodyguard and martial artist Peter Consterdine who acknowledged this book as a blueprint for the future of security. 'Old school' enabled me to put my days as a doorman to a peaceful end. It was my final release from, and goodbye to, the life that at one time was the only life I knew.

Its sales took off like crazy and has made me a household name within the door-world without me having to kill anyone, sell drugs, or be villainous.

All I did was write a good book that serves as a doorman's bible. Wherever you go in the country you will be seeing aspects of my book being adhered to within the security field.

From then on I could only go onwards and upwards so put my fingers to the computer keyboard once more and wrote my third book, '*What makes tough guys tough?*

This book really opened up the floodgates within the martial arts world for me and for its foreword I wanted someone really special.

My old pal Dave Turton who represents everything that I respect within the martial arts. 'Tough guys' way outsells my previous titles and has firmly established me as an author within my own field and has also crossed many other boundaries. It also had the knock-on effect of opening other doors such as my time being booked to teach self-protection courses at a level that I, as the bullied child, could never conceive.

It is so hard for me to get to grips with the fact that I am now a much sought after Self-protection instruction and adviser. I have my own column in Britain's leading martial arts magazine and an author with six books on sale, own my own house yet a few years ago was homeless, had my dream car imported direct from Japan and was able to pay the thousands of pounds for it in cash. I also now have my own book publishing company specialising in Self Protection with 15 current titles on sale. My dream to remove myself from door-work 5 years ago was the right choice for me. Although all the materialistic side of things make life more comfortable, I would swap it all for just one more day with my mother. The most important thing that I have gained and earned is the respect and love of my children. They adore me and that is priceless.

All this I owe directly and indirectly to the martial arts.

My study and analysis of the martial arts over two decades has structured my life greatly. Without martial arts in my life I would certainly have been a much different person than you see now. Although I owe my existence to my mother and my street life awareness to my stepfather Peter, I believe that it was the martial arts that stopped me from pursuing a life of crime. It gave me an interest and burning desire for more knowledge.

My family tree shows nothing but a trail of ignorance. Be it from my mother's Scottish line, my biological father's Irish line, or my

Stepfather's English line. This does not refer to other people from these countries; I am talking purely about my own family. You will not find any college professors, writers, doctors, or astronauts in my family tree. I hope to break the chain of ignorance and give my children and those that come after me a better start in life, hopefully seeing that there are other things out there that you can work towards.

I left school with only a basic raw understanding of the English language yet sit here now writing my next book. I wanted to write songs so I taught myself to play music. I'm no Mozart but I learnt enough to get paid £60,000 to do it.

I wanted to learn to fight so I trained and trained and trained until I became pretty good at it. I was awarded my 5th Dan black belt from Peter Consterdine and Geoff Thompson of the British Combat Association then in June 2001 I received my 6th Dan from Dave Turton of the Self Defence Federation. It took me 27 years to get there but a well worthwhile journey. Almost every single friend that I have has come into my circle in someway connected to the martial arts. I have a lot to thank the arts for.

Over the years I have done my fair share of knocking traditional martial arts because of the way I felt about them at the time. I am guilty of biting the hand that feeds me. On reflection I now realise that each and every individual art has its place in the martial arts world and every one of them has something to offer. Just because I personally do not like certain aspects of an art it does not mean that it is crap!

It simply means that I do not like that particular method of training for myself, however the stuff I choose to leave behind may be very useful to others.

I used to knock different arts because of their ineffectiveness in reality and I practically buried Karate because I felt so cheated by all the time I put in with no results. I can now see that it wasn't karate that was ineffective, in reality, it was a combination of my own inability and the misinformation that had been passed on to me. Just like if you gave me a box of tools and some planks of wood and asked me to build a chest of drawers. You would end up with a useless item containing many weak links and falling apart at the seams. Give the same tools and wood to a carpenter and you will get a better job. Give the same job to a furniture maker and you will have the real McCoy. The same thing happened to me in karate. I was given the tools and the wood (Wood being my black belt) but did not have the correct information to make it work properly for me. I was basically being taught to build furniture by people who had never built furniture themselves. I am not talking here about my first introduction into karate and the martial arts, I am referring more towards my days as a new black belt and the time put in to get there.

The combination of my own fighting ability and guidance from my mentors Geoff Britton, and my Stepfather Peter began to change all that. It took many years study to pass and a lot of 'life' lessons to be learned until I reached my final conclusion of the martial arts.

I can now take practically any technique from any art and make it work for me. The reason for this is that I have come to learn that it's not what you use that's the most important factor of success, its how you apply or use something that makes it work.

Whilst working as a doorman I began writing my first book *'Dogs don't*

My friend and biggest fan – Micky B

know kung fu' This book took me two years to write and 20 years of research. This was an important landmark within my family because it was the first actual tangible thing that anyone could see of my attempt to break the chain of ignorance within my family tree. All I personally had to look back on was criminals, alcoholics, child molesters, drug addicts, domestic violence and so on. I have protected my children from these visions and the worst thing that they could possibly see in me is that I was a bouncer. However in another year or two that guise will have been forgotten and all they will see is writer, author, black belt, teacher and so on. This will give them something from which to work on and hopefully improve on so that they can become teachers, doctors, lawyers or whatever profession they choose rather that accepting alcohol, drugs, and villainy as an acceptable way of life. All children need heroes and its better that they have a good guy to look up to rather than a bad guy.

Another thing I did whilst working as a doorman was that I put myself through college. I enrolled into college as a mature student and achieved my City & Guilds 730 'Further and Adult education teaching certificate' parts 1 & 2.

That's not bad going for someone who left school without a single qualification.

It took me 15 years to push myself back into the role of school student but I did it, and boy was it hard. To your average academic it would be a piece of piss but I wasn't your average academic.

I was a divvy doorman who so badly wanted to change the direction my life was going in that I did something about it.

I went on from there to study teaching and gained my Cert. ed. 'Certificate in Post Compulsory Education and training,' from Greenwich University. Then my D32 D33 NVQ4 Assessors award. The teacher training was my admittance and acceptance to myself that there must be a better way in which to teach people. As I did with the martial arts, I went back to class and learn how to teach. I do not know of another martial arts instructor that has done this. I'm sure that there must be some that have, but I personally don't know of any. Most martial arts instructors have acquired invisible teaching skills along with all the other hidden talents that a black belt seems to give them. They become fitness experts, champion street-fighters, philosophers, bouncers, bodyguards and the like. All this from passing a black belt exam, Wow!

I wanted to learn how to teach like teachers do then apply this to the fighting arts. I am now a pretty good teacher even if I do say so myself. I have gone from an uneducated schoolboy to a certified teacher and it's something I'm very proud of. One very important thing that I have learnt from teacher training is that everybody learns at different levels in different ways. This is something badly missing from the martial arts.

In martial arts classes everybody is taught exactly the same things in the same way. Those individuals that find it hard to learn by that method will naturally feel inadequate and will give up. However if you treat people as individuals and then find a way to teach them something you will make great progress.

An example of this would be something like a left jab.

One student may understand the body mechanics just by watching you do it whereas another may understand it better if you use diagrams. You may even have to video them so that they can see themselves, or they may just need a heavy bag for them to hit. At the end of the day each student will still end up with a jab but may have learnt it in different ways. So that's the most important discovery I made in my teacher training. This has also helped me within other areas of life when I'm trying to explain something to someone. I always try to talk to someone at the level of understanding that they portray to me.

I don't talk to a ten year old about corruption within door-work, just as I wouldn't talk to a professor about how many pints I can drink on a Saturday night.

I think that I have done a pretty good job of breaking the chain, so far. So where does that leave me now?

Well I intend to carry on writing my books and hope that I can stop before I become guilty of writing about nothing. I have half a dozen books sitting in line waiting for me to edit and release and I also have my own book publishing company where I release books for other people. So if you are a budding author write to me.

Me now –still unforgiving, still ugly

I have seen violence of all shapes and forms and can honestly say that it sickens me. As a child I was bullied and forced into fighting. Fighting eventually became a pastime for me, something that I did for enjoyment just as some enjoy fighting in the ring. I then progressed onto using violence as a means of dealing with the filth of this world that sell drugs to youngsters. From there I progressed to anti violence as a doorman that was doing my bit to eradicate drugs and violence from pubs and clubs until finally ending up in my rocking chair writing about violence in some shape or form.

So the big question here is what do I do if confronted with violence now?

Do I forgive my attacker for the error of his ways and try to understand why he is acting like this, or do I beat his brains in?

Personally I would be a very happy man if I never get into another fight in my life. It really pisses me off when someone forces me along the path of violence, but I am also not able to turn the other cheek and find forgiveness. If somebody hurts a member of my family I would not think twice about unleashing every nasty form of painful application that I know of, don't get me wrong, I'm not a monster. If you spill my beer in a pub I would not consider that to be something worth fighting for. If you road rage me I still do not consider it something worth fighting for. However change the scenario and road rage me whilst I have children in the car, putting their lives at risk, then I would not hesitate to come tearing through your street door at 5am and break every bone in your body before you even wipe the sleep from your eyes. I've done a few times when it was deserved.

My good friend Geoff Thompson has said to me that I still have a lot of anger built up inside of me that still makes me want to deal with violence in a violent way. Maybe he is right, I don't know! But what I do know is that for me it feels much worse if I let an injustice to my loved ones go un-dealt with. I do not feel any sense of personal satisfaction unless I repay the crime with plenty of added interest.

The low life that hurt innocent men, women, children, the elderly, babies, and animals all deserve a punishment much worse than they have committed themselves.

My late mother always said to me, *'Two wrongs don't make a right'* which is very true, but I say revenge sure does feel good.

If society adequately dealt with the bad guys who cause pain and suffering to their victims then maybe I would not be preaching revenge feels good, but until then I can only be honest about the way in which I feel.

I am going to do my best to avoid violence from this day on and hope that I can deal with any future problems in a manner that will give a satisfactory result without the use of violence. But at the end of the day I am still teaching self protection, and part and parcel of that is teaching one human being how to hurt another. Even if it is taught under the guise of self-defence, I am still teaching people how to seriously hurt and maim anybody that attacks them.

Society loves to hear stories of justice against wrongdoing and happily accepts revenge where justice doesn't suffice.

I don't think that anyone deep down wants to turn the other cheek and accept violence as a hazard of life but not that many are prepared to deal with it. When people like myself do the dirty work for them they are pleased but in the shadows you also hear outcry about how the law should not be taken into our own hands.

I would love to be able to tell my family, friends and students that you can deal with vicious and nasty people purely on a non-violent manner, but if I did so I would be telling a lie.

Everybody in life applies a value to something and according to that value they react accordingly if somebody else devalues it.

I put a much higher value on the safety of a loved one than I do on an insult to my pride, whereas another person may apply the same value to the paintwork of their car that I do to the protection of my life.

I try to categorise thing as simple, serious or life threatening and act accordingly. Hopefully this will give me a more peaceful future. If you would like to read more about my methods of dealing with situations at these three levels read a copy of my book 'Dogs don't know kung fu'.

I also suggest you check out 'Pre-emptive strikes – for winning fights' and of course my newest release 'No one fears when angry! – the psychology of confrontation.'

On a Final Note!

Well there you have it. One person's story and how it was inter-linked with many other people's lives. I now realise that every individual's actions affect other people in many different ways and that I must be careful as to how my actions affect others. For every action there is a reaction.

Since my Mother and Stepfather Peter died I have had nobody senior to me to share my troubles with, no adviser to help me through and no one to push me into different choices. I have had to do it all on my own and in turn have become an adviser to my young ones and others around me.

Life has been hard for me and I certainly think that I have taken a fair beating with the unlucky stick, but on reflection I can now see that all my experiences have given me a vast variety of ways in which to deal with new problems as they arise.

As for my family and their problems? I still have a dedicated phone line installed in my home just for family to ring me on. As soon as that phone rings I am already lacing up my trainers and it can be guaranteed that I am going to have to pay a visit to someone. I absolutely hate it and still get that sickly adrenal rush that I used to get on the door when someone shouts, 'It's kicked off', but family are family, what can you do?

So when you look at someone be it a bouncer, boxer, doctor, inmate, or astronaut and you either admire or dislike that person. Have a good think as to what path they travelled to get where they are today. What is it that makes them act the way they do and make the choices that they make now. Even ask them, *'What's your story'*, and you will find that there is a different answer than, 'Morning Glory', from some of the most interesting people that you will ever come across in your life.

So, would I live my life again as I have? With hindsight yes, but without it no. I am not proud of my life. It's just as it is.

All I can now do is try making the rest of my time that hopefully I have left, an improvement on the past. However I have matured, evolved and reinvented myself to become a shining example that goodness can come from bad.

If I could pass on just a few words of wisdom to my own children, or anyone else it would be that

'People only treat you the way you 'allow' them to'
and that,
'Anybody can do anybody - you just have to find a way'

Thank you for reading my story. I will look forward to reading yours.

Jamie

Jan 2002

Adverts
Will the reader please note that the following advertising section of the book is included to let you know of other Self Protection related merchandise.

You the reader, have not been charged for the printing or paper used in this section. The cost for this has been absorbed by New Breed Publishing.

The price that you have paid for this publication is for the knowledge, information and advice given by Jamie O'Keefe throughout the rest of this book.

Thank you

INNER LONDON EDUCATION AUTHORITY

SEBRIGHT JUNIOR SCHOOL
Maidstone Street, E.2

P

REPORT FOR YEAR ENDING *July*. 19 71

Name *James O'Keefe*.　　Number in Class 27

Class 3　　Position in Class

SUBJECT	ASSESSMENT	REMARKS
ENGLISH Reading	Excellent	James has done some
Comprehension	Excellent	excellent creative writing.
Composition	Excellent	
Spelling	Good, but sometimes careless.	
~~ARITHMETIC~~ Mathematics	Improving. After a slow start James is settling down to Mathematics	
HISTORY ⎫ GEOGRAPHY ⎬ SCIENCE ⎭	Very good. Shows keen interest and good general knowledge	
ART	Very good.	
WRITING	Can be very good, but sometimes careless.	
HANDWORK	Very good.	
OTHER SUBJECTS		

Religious Knowledge *Good*.　　Attendance *26 Absences out of a possible 280*

GENERAL REPORT *After a slow start James has shown that he is capable of some fine work. He seems to have settled into the class now.*

M. J. Gray. Class Master ~~Class Mistress~~　　*J E Evans* Head Master ~~Head Mistress~~

**My school report from Seabright Junior School, East London
When I was 10 years old**
'James has done some excellent creative writing'

Kevin O'Hagans latest book
BAD TO THE BONE
Exploring the many facets of violence
and aggressive behaviour

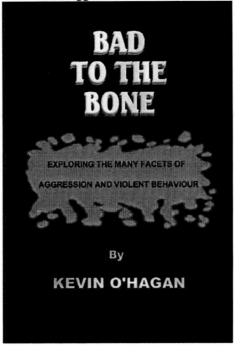

£14 inc Post and packing
from New Breed, Po box 511
Dagenham, Essex RM9 5DN

IN YOUR FACE - 'CLOSE QUARTER FIGHTING'
By Kevin O'Hagan
Available now at £14 inc p&p from NEW BREED
WWW.NEWBREEDBOOKS.CO.UK

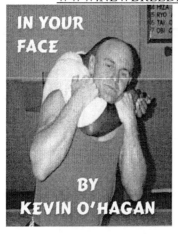

Close Quarter Fighting
£14 inc p&p from New Breed

Future NEW BREED publications
By Steve Richards

PREVENT YOURSELF FROM BECOMING A VICTIM
'Dogs don't know Kung Fu'
A guide to Female Self Protection
By Jamie O'Keefe **£14** including post & packing

Never before has Female Self Protection used this innovative approach to pose questions like. Why do Rapist's Rape? Why are Women abused? Why do Stalkers Stalk? This book takes a look at all Simple, Serious, and Life threatening aspects of Self Protection that concern us daily, along with **PREVENTION** of Child abuse and Child Abduction, Emotional cruelty, Telephone abuse, Road rage, Muggers, Date rape, Weapon attacks, Female abduction, Sexual Assault & Rape, Self defence law, and what it will allow us to do to protect ourselves, plus much more. With over 46,500 words, 77 pictures and 200 printed pages 'Dog's Don't Know Kung fu' is a no nonsense approach to women's self defence. It covers many realistic scenarios involving Children's abduction as well as typical attacks on women. Besides quoting actual events, the book explains how to avoid trouble and how you should react if you get into a situation.

This book is a 'must read' for all women and parents.

It is also important for teenage women, but, due to some of its graphic depiction's of certain incidences, parents should read it first and decide if it's suitable for their child.

WWW.NEWBREEDBOOKS.CO.UK

191

192

The Secret Domain

WHAT MAKES

TOUGH GUYS
TOUGH
The Secret Domain
by Jamie O'Keefe

EXCLUSIVE
INTERVIEW
WITH
ROY SHAW
ON
WHAT MAKES
TOUGH GUYS
TOUGH

Written by Jamie O'Keefe

Jamie O'Keefe has interviewed key figures from boxing, martial arts, self-protection, bodyguards, doorwork, military, streetfighting and so on. Asking questions that others were too polite to ask but secretly wanted to know the answers.

Interviews include prize-fighter **Roy Shaw**, also **Peter Consterdine, Geoff Thompson,** and **Dave Turton** from the countries leading self-protection organisations 'The British Combat Association' and the 'Self Defence Federation.' Along with Boxing heroes **Dave 'Boy' Green** and East London's former Commonwealth Champion **'Mo Hussein.' Plus unsung heroes from the world of Bouncers, Foreign Legion, Streetfighters, and more.**

This book also exposes the Secret Domain, which answers the question 'What makes tough guys tough.'

Find out what some of the toughest guys on the planet have to say about 'What makes tough guys tough' and how they would turn you into a tough guy.

Available from NEW BREED at £14 inc p&p

WWW.NEWBREEDBOOKS.CO.UK

**At last a book just for
the smaller person**
Ever wondered how the little guys
Manage to beat the big guys?
Wonder no more!

I THOUGHT YOUD BE
BIGGER

£14 from New Breed

**FROM BULLIED TO BLACK BELT
BY SIMON MORRELL
from
New Breed, Po box 511
Dagenham, Essex RM9 5DN
£14 inc Post and packing**

Pre-emptive strikes
for winning fights
'The alternative to grappling'

by
Jamie O'Keefe
£14 inc P&P
from
New Breed
Po Box 511
Dagenham, Essex RM9 5DN

AVAILABLE NOW

BY
Kevin O'Hagan
£14 inc Post and packing
from
New Breed, Po box 511
Dagenham, Essex RM9 5DN

The latest book by Jamie O'Keefe
'NO ONE FEARS WHEN ANGRY'
The psychology of confrontation

Everything you need to know about dealing with ANGER

Please feel free to review any of our books on
www.amazon.co.uk

Why not also look at the dedicated websites of the
New Breed Authors
Jamie O'Keefe
www.newbreedbooks.co.uk

Kevin O'Hagan
www.bristolgoshinjutsu.com

Alan Charlton
www.spa.ukf.net

Steve Richards
www.renaissance-academy.com

Simon Morrell
www.frombulliedtoblackbelt.net

www.blackbeltkarate.net

The new book by **Alan Charlton**
NEW BREED PUBLISHING £14 inc P&P
PO BOX 511, DAGENHAM, ESSEX
RM9 5DN

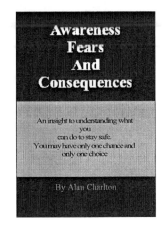

Awareness
Fears
And
Consequences

An insight to understanding what
you
can do to stay safe.
You may have only one chance and
only one choice

By Alan Charlton

Steve Richards

The memoirs of a Kung Fu cop
Part One – Beginnings
Topics include death, sex, serious injury, romance, pathos, tragedy, humour, riots, IRA, drugs, bouncers, getting planted and just about everything you'd expect from a Scouse Copper over 13 years front line duty. Also, loads on traditional Chinese arts from within their closed shop community, and how they worked or otherwise under real pressure. The post-police part of the book covers his martial arts and other careers since that time

£14 inclusive of P&P Available from
New Breed, Po box 511, Dagenham, Essex RM9 5DN

SAFETY FIRST (UK)
Training Available
Our current list of training programmes and services:

Accredited Conflict Management & Personal Safety Training

Open College Network Accreditations

- *Introduction to Personal Safety - Level 1 **
- *Personal Safety - Self Protection - Level 2 **
- *Staff Personal Safety Awareness - Level 2 **
- *Conflict Management Instructor's Training - Level 3 **

Edexcel Foundation/BTEC Qualification

- *Conflict Management Instruction - BTEC Advanced Award **

Additional 1 & 2 Days Short Courses

- *Recognition and Resolution of Conflict at Work*
- *Staff Personal Safety Awareness Training*
- *Personal Safety/Self Defence Training*
- *Stress Management & Relaxation*

Specialist Training & Services

- *Personal Security*
- *Control & Restraint Training*
- *Close Quarter Defence Techniques Training*

Available as Distance Learning Programmes

For further information regarding our training programmes or services, please contact:

Safety First (UK), 2 Lansdowne Row, Berkeley Square, Mayfair, London, W1J 6HL.
E-mail: personalsafetyfirst@hotmail.com
Telephone: 0207 306 3399
© **Safety First (UK)**